LIGHTS
OUT

LIGHTS OUT

Pride, Delusion,

and the Fall of

General Electric

THOMAS GRYTA AND **TED MANN**

Houghton Mifflin Harcourt
BOSTON NEW YORK
2020

For information about permission to reproduce selections from this book, write
to trade.permissions@hmhco.com or to Permissions, Houghton Mifflin Harcourt
Publishing Company, 3 Park Avenue, 19th Floor, New York, New York 10016.

hmhbooks.com

Library of Congress Cataloging-in-Publication Data is available.
ISBN 978-0-358-25041-8

Book design by Margaret Rosewitz

Printed in the United States of America
DOC 10 9 8 7 6 5 4 3 2 1

To our wives, Virginia Gryta and Annie Karni

CONTENTS

LIGHTS
OUT

1

OFF A CLIFF

Schenectady, New York, 2017

J OHN FLANNERY PULLED into the little city on the Mohawk River in late July with numbers on his mind, passing beneath what had once been the largest electrified sign in the world. The storied logo had been surpassed by taller structures long ago, but it still glowed from atop the mammoth brick shoebox of Building 37 as Flannery passed through the gates and into the birthplace and spiritual home of the General Electric Company.

Schenectady—the electric city—had been home to the Edison Machine Works. It was there that GE was incorporated in 1892, assembled by bankers out of the nonperforming attempts of inventors to translate their brightest ideas into business.

What had blossomed in Schenectady was the stuff of cultural legend —inventions, manufacture, mass production, unstoppable growth— but there was a sense of cavity now to the giant old industrial grounds. More than forty thousand men and women had worked there at GE's peak. It was a tenth of that size by 2017.

History, however, wasn't Flannery's mission. His focus was inside. Schenectady now was the headquarters of GE Power, the largest and oldest division of what had long been America's most famous conglomerate.

And John Flannery, just weeks away from taking over leadership of

the most famous C-suite in American business, had come to see GE Power leaders on their home turf and to take stock of the heart of the company he was about to lead.

Flannery was like a president-elect, the incoming chief executive officer of one of the most famous and well-respected companies on earth. Just ten other men had held the job he was preparing to take on. With the formal and official announcement made, Flannery was organizing his team and girding himself to take on the hardest challenge of his professional life.

To an outside eye, nothing more than another orderly and carefully planned corporate succession was under way, one as smooth as those GE had prided itself on in the past. Flannery's predecessor, Jeff Immelt, was overseeing a peaceful transfer of power before making way for a new manager to rise up from GE's ranks and carry on the company's traditions for another decade or so.

But appearances were deceiving. What was actually unfolding behind the scenes at GE was dysfunction tending toward chaos and a confrontation with the past that was mere weeks from spilling into public view. Beneath the placid surface, GE was in total disarray.

Flannery had barely had a moment to reflect after the company's board tapped him to be the new CEO. That whirlwind weekend opened into a week of press conferences, media interviews, internal town hall–style meetings, and multiple executive briefings — all compulsory steps in the process of preparing Flannery to take the reins of one of the world's biggest and most closely watched companies.

And prepare he would. Flannery was a voracious reader, his wide-ranging tastes reflected in his conversations, but he didn't ooze arrogance as some corporate chieftains do. He was a man constantly reexamining himself, his curiosity often reflecting inward as he reviewed his past calculations and decisions like an analyst poring over a slide of film.

That same banker's instinct to endlessly search for new angles and to weigh his options, to crunch and recrunch every number, flowed from the same quietly adventurous source that led him to venture down dirt roads with his wife in exotic locales. He was a hunter — for killer deals and hidden risks, for undiscovered roadside taverns serving lunch.

Flannery stood a little under six feet tall. He was slightly stout and usually wore dark suits that reflected his finance roots. He wasn't shy,

but he wasn't one to work a room either, in contrast to some of those he had outmaneuvered to win the new job. Some GE executives glad-handed as aggressively as candidates for Congress. Flannery was self-deprecating, though possessed of a disarming confidence. At company events for investors and the press, a wry smile often played at the corners of his mouth, a contrast with the brand of GE earnestness exuded by his colleagues.

He would have to adjust to his new stature, however, as he took on a kinglike role in a company that took itself just as seriously as a kingdom. But he hadn't yet. Like a world leader, Flannery would need to get used to being whisked from one place to another, having a full security detail, and constantly having a car, plane, or helicopter waiting to deliver him to the next stop. In the hours before his predecessor, Immelt, arrived for one of his innumerable visits to GE facilities all over the world, telltale trappings would begin to appear at his destination: hard candies and a plentiful supply of his favorite diet soda appeared on shelves in conference rooms, always there before anyone in Immelt's entourage even had to ask. Rumor among stock analysts held that he flew with his own treadmill, lest the hotel gym prove insufficient. No one knew yet what soda John Flannery liked. And Flannery certainly wasn't used to any of this.

When he walked to the elevator bank in GE's Boston office, an assistant immediately scrambled over to hit the Down button, apologizing that the elevator hadn't been waiting for him. Flannery appreciated the effort, but the royal treatment seemed a bit overboard and he told people so.

Working closely with Immelt, he had seen these trappings before, but now, Flannery told people, he found them slightly suffocating, and occasionally a little silly. Nevertheless, this was the job. With its vastness came complexity, the crammed agenda, the aides and meetings, the planes and guards. Although he wasn't going to be able to ditch the entourage anytime soon, at least he had a good excuse to get away from headquarters on trips like this one to Schenectady. No one challenged his need to get acquainted with the details of the business.

A road trip was also good for clearing the head. John Flannery had been announced as the new CEO on June 12. His father, John, a retired banker in West Hartford, Connecticut, had died just twelve days later.

It was some comfort to the younger Flannery that he had been able to share the news of his promotion, and that his father had lived to see his son tapped to land one of the most prominent jobs in American business. Still, his dad's passing hurt. Loss shaded his satisfaction and pride.

Flannery had outperformed three rival GE executives to win the contest to succeed Immelt. The latter had served sixteen years as CEO, but had shown very little outward sign that he was ready to retire, even to those aiming to succeed him, until just before his departure was announced.

Flannery was a finance whiz, a veteran of GE's large lending business, which made him a dark horse at first in the race for CEO, given the company's traditional reverence for its industrial businesses and their leaders. But the board knew that GE needed a fresh assessment. Immelt's strategy was stuck in the mud, and his supremely optimistic mantras simply weren't falling to the bottom line.

Even from his first moments on the job, Flannery wanted his tenure to be defined not just by what he would do but by what he wouldn't do. He wouldn't just kick the can down the road. Instead, he would rip off Band-Aids and expose some of the festering ailments within the company to fresh air and sunshine. That meant dealing with the truth, no matter how harsh it might be and no matter the consequences. Flannery would be brutally honest, even though, as he was well aware, that would mean changing the company's tone. Under Immelt, there had been a buzzy, vague, optimistic spin that not only often failed to hold up under scrutiny but had eroded GE's credibility with Wall Street and its workers alike.

Flannery knew that his tenure as CEO would last at least a few years. The GE board of directors would give him some time. But he also had to set a new tone from the start and get started right away on changing what needed changing, even purging where needed. And he knew there was plenty to purge.

Flannery had taken to uttering a new mantra around the company's shiny new offices in Boston: "No more success theater."

Now, with just weeks to go before officially starting his new job, he was working around the clock to assess the company. Like any CEO, Flannery wanted to survey his new territory so that he could make decisions on strategy and assess performance based on his own firsthand

observations of the company—its factories, its offices, its profit-and-loss statements, its debts. He had already met with more than one hundred investors and financial analysts in the previous weeks. Now he was visiting the GE Power division, before heading to the GE Aviation facility in Cincinnati.

He wasn't alone, physically or emotionally. In fact, finding a way to be alone in order to think was sometimes difficult. And it was also sometimes difficult to see what he needed to see. GE was a siloed organization, in contrast to the image it presented to outsiders. By the standards of a normal business career, Flannery had been all over the GE corporate map, making stops in financial services, running operations in Asia, India, and Latin America. He had run the business development team when it bought one of GE Power's biggest global competitors.

But that wasn't the same as having worked at Power itself.

He didn't know the intricacies of the markets, the products, the cycles, or the people. He didn't know the myriad ways in which executives in that unit had adjusted accounting, calculated estimates, or weighed risks before they reported their figures up to the places in the company where an executive like Flannery could have seen them. Even for a GE lifer, the only way to know what the Power unit was actually up to was just to show up at its headquarters and look around.

So there they were: Sitting in a conference room in the heart of the Power business. Flannery and his team sitting on one side of a long table with the Power management group facing them. As they discussed the business, the expressions of the two sides began to look vastly different.

Flannery was comfortable with numbers, especially financial statements, and that was where he began his search in Schenectady. It didn't take long to see the problem: as Flannery paged through the financials, he realized that GE Power had somehow run out of cash. This discovery wasn't just shocking—it was unthinkable. GE's largest industrial business was stretched thin. Its profits, on close examination, seemed to exist mostly on paper. Years of pro forma adjustments had given the appearance of a business that was turning decent profits selling power turbines and the services that kept them running, but in fact there was relatively little actual money coming in the door from customers. Even worse, Power was building inventory—making more of its huge, expensive machines—even as the global market for turbines was slowing. "It

was like they drove off a cliff," Flannery later told an observer, "and there were no skid marks."

The gas turbines that made up the core of GE's business were essentially cousins of their aircraft engines: enormous spinning rotors moving equally titanic generators not that different from those first installed in the early days of Edison in Lower Manhattan. And no one made more turbines — the massive machines at the center of power plants — than General Electric, whose equipment still generates about one-third of the world's electrical power.

The GE Power unit that Flannery inspected in 2017 was under new management. Its leader of the past dozen years, another GE lifer named Steve Bolze, had put in for his retirement soon after losing out to Flannery in the contest to be GE's next CEO.

Bolze's exit wasn't surprising once the succession contest had ended. But the decision to deprive him of the GE crown was shocking for some — primarily for Bolze himself. He had the tall, square-jawed looks and charisma of an Ivy League quarterback or TV weatherman and had been telling people he had the job just days before the GE board made its decision.

On paper, Bolze had looked like an obvious candidate to succeed Immelt. He ran the biggest division and had helped close GE's biggest-ever takeover, and his résumé looked like a decades-long campaign for the job. Like most ambitious GE executives, Bolze had made stops at other businesses across the conglomerate. He had run GE Healthcare's overseas operations. He had worked out of headquarters on the deals team that did mergers and acquisitions. He had allies on the board. And Bolze had presided over enough growth in his twelve years running Power to justify a place on the short list for CEO.

But Flannery's review of the GE Power books now raised the question of just how real that growth had been. Had Bolze thought that the condition of this business was sustainable? Did he think that the power market could come around? Or was he planning to deal with GE Power once he was GE's CEO and not beforehand, when reporting bad news — especially to a CEO like Jeff Immelt — would have killed his chances at the top job?

Or was he unaware of the true condition of the division, hardly a forgivable offense, until it was too late? The pressure to perform inside GE

is omnipresent, and missed goals can be fatal, a tradition true at all levels of the company. Even as head of the division, Bolze didn't necessarily know how his underlings got to the finish line and it didn't really matter. Such details are fixable at GE, but missing the financial target for your business causes irreparable damage.

In any event, with Bolze now gone, Flannery needed to get his arms around the Power business and fast. He needed Jeff Bornstein, who came along for the trip. The chief financial officer was in some respects Bolze's opposite: short, punchy, funny, and ensconced at the top of an intracompany network of financial chiefs that pervaded every last tentacle of the corporation. Like Bolze, Bornstein was another of the finalists Flannery had beaten out to become CEO. Unlike Bolze, Bornstein had stayed on, pledging to help the new CEO get GE back into shape.

Sitting in Schenectady, surrounded by newly crunched copies of Power's numbers, Flannery's mind reeled. The more he dug into the numbers, the more the problems grew and worsened as they built on each other. Not only was the Power business poorly positioned for any turn that might happen in the market, but it didn't have the cash to fix itself. To GE investors, Power seemed to have been making its numbers and putting up solid profits. But those were illusory. The accounting tricks that looked like profits were actually just borrowing from the company's future earnings to cover up problems in the present.

Power had sold service guarantees to many of its customers that extended out for decades. By tweaking its estimate of the future cost of fulfilling those contracts, it could report boosts to its profit as needed. Flannery shook his head; he couldn't believe that a major GE division had dug itself such a deep hole.

The world wasn't going green overnight, but as more alternative power sources came online, natural gas turbines were used less and needed less frequent service, which was the real cash cow of the Power division.

In the coming weeks and months, demand continued to drop for gas-fired power turbines. Competition from wind and solar had been growing for years and was only getting fiercer. Meanwhile, the division was sitting on too much unsold inventory, which was just capital that the company couldn't access without selling the equipment. And selling it wouldn't be an option as the market continued to sour. Somehow a po-

tentially fatal spiral in GE's biggest business had been gathering force out of sight of GE's generously paid board of directors and many of its top executives, save a small circle that included Bolze and Immelt himself.

Now, just weeks before becoming CEO of the General Electric Company, John Flannery sat in Schenectady watching disaster approach on the horizon. Combing through the books of the sprawling conglomerate in preparation for taking on the job of his life, Flannery looked into the biggest and most important industrial business of them all — the unit that was the reason for GE's existence — and found a deep, empty hole where there should have been cash.

The reported profits were aspirational, if not fraudulent. And the accounting devices that had hidden this disarray from the public were beginning to fail. Thirty years into a career at America's most iconic company, John Flannery had reached the pinnacle. But now the whole corporation was about to plunge into the abyss.

Flannery was understated, even in panic. Still, a palm upturned, an eyebrow raised, made the incoming CEO's thought clear to anyone in the room as he turned from the charts to the financial chief he had known for two decades.

"Did you fucking know about this?"

2

THE MEATBALL

Boston, 2019

THE LOGO IS one of the most identifiable in the world: a circle, white on a bed of blue, with four subtle, scalloped winglets, suggesting the blades of a midcentury tabletop fan. In the center, in a typeface that has been altered only slightly over the decades, are two entwined cursive letters: GE.

At General Electric, the corporate logo is officially known as the Monogram. The old-timers have a more affectionate name for it. They call it "the meatball."

For more than a century, the conglomerate has stamped its distinctive logo on a dizzying range of objects that reflect the businesses that it has pioneered, acquired, or briefly dabbled in. The meatball can be found on jet engines, ultrasound scanners, wind turbines, televisions, commercial loan agreements, clock radios, toasters, nuclear reactors, lightbulbs, security systems, tubes of silicone caulk, wing-mounted rotary cannons, locomotives, and washing machines. One estimate pegged the value of the GE brand represented by the Monogram at nearly $30 billion.

Since its founding in 1892, General Electric has been more than a corporation. It has been an American institution. For decades, it was a winning lottery ticket for its hundreds of thousands of employees and a safe bet for shareholders. For its executives, it was an elite business education, and for some a path to enormous riches. GE electrified Amer-

ica, powered its biggest machines, and became integrated into American society as few companies ever did. It was so large that it was given the same brand of financial credit and trust as the US government itself. GE melded Thomas Edison's workbench with J. P. Morgan's financial might to create a juggernaut that, in powering the nation's middle class, its military might, and its explosion of financial wealth, marched in step with the rise of modern America.

GE grew as the nation grew, managed to evolve with the times, and entered the twenty-first century stronger than ever. At its peak in 2000, General Electric was the most valuable company in America, worth almost $600 billion, and had business lines that sprawled across boundaries to touch vast swaths of life in the developed world.

GE's industrial machinery and consumer goods electrified the power grid and lit American homes and kitchens. Its engines kept aloft American fighter jets, commercial airliners around the planet, and even Air Force One. Its lenders propped up new owners of McDonald's franchises and leased out railcars carrying oil, grain, and lumber across North America. Its sonograms beamed images to expectant parents, its X-rays revealed broken bones, and its MRI machines scanned organs searching for cancers. Americans dashed to its refrigerators for snacks, then back to their couches to watch episodes of *Seinfeld* and *Friends* — also made by GE. General Electric was an industrial company, yet it seemed to sell everything.

Fewer than two decades later, the meatball can still be seen everywhere, but that GE is gone, if not unimaginable.

While still a massive operation with hundreds of facilities, GE's stock is a mere fraction of its peak value. The company is no longer a media darling or an analyst favorite, its shares are no longer counted in the Dow Jones Industrial Average, and the once-generous dividend is virtually gone. A share of GE stock was once an essential component of the beginner investor's portfolio, but is now perceived as a speculative bet; a generation ago, such a view would have bordered on market heresy.

GE's fall was fast when measured in dollars and cents and people employed. Pensioners and retirees watched money evaporate at a time in their lives when they had no way of replacing it. Thousands of employees lost their jobs, and those who remained were left with an uncertain future. Many who weren't laid off at first found themselves working out-

side the very company they had always known after GE sold off pieces of its storied history in exchange for the cash it needed just to keep itself afloat. The end of GE as we knew it, however, was set in motion long before the cards finally fell.

In one important respect, GE's demise is more profound than can be conveyed by numbers on a stock ticker or even the deeply felt pain and disappointment of individual workers and executives and their families. The collapse of a company that taught generations of American businesses what it meant to manage well raises a major question, one that is still unresolved. Just how much of the success of the many other companies that have chased and emulated GE was real? How much was a product of their imagination — and perhaps of our own?

In company lore, indestructible General Electric carries the torch of Thomas Edison, the legendarily prolific inventor. It is a connection that instantly validates the company's place in America's proud history of lucrative ingenuity. It is also, as it happens, a beneficial association for a company that sells lightbulbs.

But GE's birth, like its life, was more about money than about invention. Edison had very little to do with GE, which was the result of financial titans rolling up multiple early players in the electrical industry, as the technology's development required grand scale and, more importantly, mountains of capital.

The father of the General Electric Company proper was not Thomas Edison but J. P. Morgan, who arranged the merger of dueling rivals for just this reason. The lagging financial condition of Edison's old companies had left the great inventor with no choice but to see his former firms acquired in the deal, and with no role for himself except as a figurehead on the first GE board of directors — and a useful avatar for company public relations. But he would serve only briefly on the company's board. Edison sold his last GE shares within years of the company's formation, missing out on its meteoric growth, to finance failed mining experiments.

Besides frequent appearances in its marketing, the inventor's legacy nevertheless survived in GE research culture, which institutionalized problem-solving and made invention a team pursuit. Edison's personal presence was not nearly as critical as the inspiration he represented,

which was a hugely important factor in the company's success and its broader renown. From GE's labs flowed decades of pathbreaking research, patents, and Nobel Prizes. From its example flowed a broader heyday of private-sector research and development in the United States as other companies sought to match the success that GE had found in funding laboratories, paying scientists, and parlaying the fruits of their research into marketable goods and machines.

The company also taught its peers about the power of public relations, influence-peddling, and mythmaking. The archly rendered city of Ilium in the novels and stories of Kurt Vonnegut was a stand-in for real-life Schenectady, New York, the incorporated home of GE, which Vonnegut knew, perhaps too well, from his days spent writing copy for internal corporate communications. Vonnegut's brother was a GE scientist in Schenectady who successfully worked on developing methods to induce clouds to begin raining. Thwarted on antitrust grounds from participating in the dawn of the great American radio networks, GE nonetheless found its way into pop culture as a sponsor of live television in TV's golden age. It hired celebrity pitchmen — in particular, a film actor whose career was on the wane named Ronald Reagan — to entice consumers to buy its home goods and fire up workers with enthusiasm about the world-changing innovations being driven by GE's work. It's a truism of modern-day Silicon Valley that every startup promises that its business aims to change the world; GE beat them all to that line by at least three-quarters of a century.

GE also led its peers into a harsher and more successful battle against organized labor and in defense of the preeminence of investors, whose demand, foremost among all other concerns, was the protection of the company share price and the dividend. It was GE's vice president of labor and community relations, Lemuel Boulware, who instituted the company's aggressive posture in negotiations with its blue-collar manufacturing workforce. The company's first offer was its offer, take it or leave it. Such intransigence was so associated with GE that it became known as "Boulwarism." And Boulware's tutelage is widely credited, among those who knew him, with engineering the transformation of the company pitchman, Reagan, from a run-of-the-mill Hollywood back-lot Democrat into the first nationally electable politician of the American right. GE remembers, with pride, its role in creating Reagan.

The company surfed the fads of the American twentieth century, including the rise of the essential postwar form of American capitalism, the industrial conglomerate. The tentacles of the GE conglomerate branched into a seemingly unlimited number of areas of commerce, investment, communications, and influence. Investing in a share of GE was an obvious smart move to the mom-and-pop investors of twentieth-century America. The company was as trusted as a government bond, tied to a proud national inheritance of innovation, and paid a reliable dividend that had been cut only once, in the Great Depression. Indeed, the company was a sort of proxy for the American economy as a whole because, in addition to hiring brilliant engineers and managers, it employed hundreds of thousands of skilled tradesmen. If one sector stumbled, another industrial segment of GE could sustain the larger body of the company. It was conservative. It unapologetically made money. Its collapse was unimaginable.

To many investors, GE was just the right amount of boring: dependable as a utility, unlikely to skyrocket in price, and predictable.

The first signs of trouble were barely visible during the period of General Electric's greatest, wildest success, under the leadership of the man widely seen then as the greatest business executive of his generation. He was a tough Irishman from Massachusetts. His name was Jack Welch.

3

NEUTRON JACK

THE MOST INFLUENTIAL CEO of the twentieth century was born in the middle of the Great Depression, the son of a railroad conductor and a homemaker, neither of whom had finished high school. During the long hours when his father was away, working the trains on the Boston & Maine Railroad, the slightly built boy, John Francis Welch Jr., formed a deep bond with his sharp-tongued, ambitious mother.

It was from Grace Welch that the boy Jack picked up what would become a favorite phrase: "Don't kid yourself," she would bark. "That's the way it is."

Grace Welch taught her child self-confidence with the same style he would use as he embarked on a career in business. There were cutthroat card games at home and withering public dressing-downs, like the time she berated him in front of his school teammates for his poor sportsmanship. The result was a child brimming with self-assurance, trained for boldness, and spoiling for a chance to prove his abilities to those who might have doubted him. The small but highly competitive child ultimately became the captain of his high school ice hockey team in Salem, Massachusetts. With Grace's ever-vigilant eye on his studies, he graduated from high school and headed off to the University of Massachusetts at Amherst to study chemical engineering. The boy kept pushing and driving, earning a master's degree and a PhD in the same field at the University of Illinois.

He was not interested, however, in an academic career. Grace Welch's

terrier of a son would go into business. He was determined to make a name for himself and to make some money. In 1960, Jack Welch was hired by the Plastics division of General Electric Company.

Welch spoke in animated bursts, his voice high-pitched, raspy, and accented by his New England blue-collar roots. Although he had fought a persistent stammer since childhood, he brought the locker room bravado that had served him in Salem into the corporate world. For Jack, the world was divided into winners and everyone else. Confrontation was his steady state. Outside of work, he kept the macho sense of competition boiling with nonstop sports banter, golf, hard drinking, and the unending ribbing of both colleagues and rivals.

Jack Welch had a fiery soul that filled the room; former executives remembered how his blue eyes could pierce through a crowd when he had something important to say. One GE executive described him as "radiating noise and energy like a thundercloud." Doctorate notwithstanding, he was always known simply as Jack.

From the outset, Welch carried himself with a swagger that could tilt over into bombast. Even well into his two-decade scramble up GE's corporate ladder, Welch's genius for leadership wasn't always recognized by everyone. When then-CEO Reg Jones asked for a list of possible successors in the late 1970s, he was given a list that didn't even include Welch.

Jones was the quintessential CEO in the traditional GE image: a lean, highly admired Englishman who was twice asked to join President Jimmy Carter's cabinet. Jones was modest to the point of self-effacing. According to company lore, he had once introduced himself so circumspectly to the mayor of the city of Bridgeport, which adjoined suburban Fairfield, where Jones had moved the headquarters from New York City, that the mayor had at first thought that Jones was the manager in charge of a local Connecticut office. When Jones gently clarified, the mayor exclaimed, "You run the whole fucking company?!"

Welch, a young man in a hurry even by GE standards, had left no one with an impression that he was less important than he was. Style aside, Jones was impressed with Welch's strategic planning, even if he lacked the polish and reserve of a typical "GE guy." The CEO added Welch to the succession short list himself.

The company's shares performed miserably under Jones, but he was still considered one of America's leading business executives, both of

GE and of his era. GE's stock woes were largely explained away as stemming from macroeconomic forces outside the company's control. What was well within the company's power, however, was solid. GE's revenue and profits more than doubled during Jones's nine years as chief executive, and amid the broader economic turmoil of the late 1970s, the company maintained its steady dividend and huge base of individual stockholders.

Meanwhile, the perpetually driving Welch continued to progress upward through the executive ranks. Jones had first encountered Welch after the young man left Plastics to join GE's strategic planning operations, a team at headquarters that laid out multiyear plans for business development, acquisitions, and streamlining of the company's portfolio. (The unit was the pride of Jones's operation, but Welch found it stifling.) From the planning office, Welch was promoted again, to run GE's consumer products business, churning out radios and toasters for the American mass market.

The consumer products job turned out to be only a stepping-stone. In just twenty years, Jack Welch was named the ninth chief executive officer in the history of GE.

When he got the top job, Welch was intent on eliminating bureaucracy in order to reduce excess costs. Welch was implementing and expounding the corporate philosophies that would define the 1980s and '90s and make him a corporate celebrity. Welch's core mission was to attack complexity, ripping out layers of bureaucracy that had built up inside the company and making the massive company more nimble. Welch would do his best to kill anything that slowed GE down.

Strategy meetings and reviews in the pre-Welch years were structured, daylong affairs, dominated by huge binders and projections for the next two decades, one GE executive recalled. Welch saw these as hopelessly lumbering exercises in a fast-changing economy that required constant updating, refining, and shifting the portfolio of businesses, products, and people. Executives like Jones had seen five-year plans for industrial businesses as conservative and essential planning tools, given the amount of investment required to develop new products and bring them to market. Welch called them "bullshit."

Welch shunned the extensive planning and shrank the team. He pushed decision-making down to the individual businesses but kept a

sharp eye on the specifics of each operation. He pushed middle managers to stop writing long memos and to lose their thick planning books. "I don't want planning. I want plans," he would say. Welch's defenders said that he had an uncanny ability to marshal detailed information from deep within GE's business lines, even as he kept an eye on the larger forces that were changing the face of global business, like outsourcing, trade policy, and the rise of Japan.

Executive teams were tasked with finding and removing layers of management. During one visit to the GE Aviation facility in Lynn, Massachusetts, Welch chatted with workers in the plant's boiler room, where he learned that the operation of the boiler room was supervised by four layers of management, a shocking discovery. That was just the sort of complexity that Welch was determined to carve out of the company.

He proselytized the new GE religion: every business should be either first or second among its competitors. Welch sold major businesses that had been central to GE's history, putting an end to the era of GE selling TVs and toasters. And he moved aggressively into new fields in search of profits, including a $6.5 billion acquisition of RCA Corporation, owner of NBC, in 1986.

The RCA deal was a huge dive into the media world that would be almost unthinkable today. The benefit of owning companies in disparate businesses would have been questioned in any era, because of the divergent skills required to churn out heavy-duty machinery in one unit and top television sitcoms in another. That diversity is even less favored today, thanks to the rise of activist investors and Wall Street's general consensus that focused businesses have higher value.

The shares of conglomerates inherently trade at a discount off the sum of their individual units, but the company's diverse set of businesses and better access to capital theoretically offer protection from volatility. In the age of cheap online trades, managed funds, and index funds, however, the so-called conglomerate discount is no longer worth it.

Under Welch, GE was changing rapidly. He famously gave a speech in his first year as CEO titled "Growing Fast in a Slow-Growth Economy." With the power of the GE brand providing credibility to his strategy, the new CEO oversaw almost one thousand acquisitions, or about four deals a month over his two decades, with a value topping $130 billion.

By 1985, with only five years on the job, Welch had spent more than

$8 billion to overhaul factories for robotics and automation. He had also seized on an important new source of profit: the financial services unit then known as GE Credit Corporation. Welch spent $10 billion at GECC, buying up property and business equipment that GECC then leased out to other companies.

It was the beginning of the assembly of one of the greatest profit engines in the history of the company, a vast portfolio of business units and properties all designed to use GE's immense, solid balance sheet not just to finance goods so its customers could buy them but to pursue financial profits for their own sake. Welch's greatest innovation, as great as his much-advertised embrace of management training and efficiency, was this embrace of finance. It would change the makeup of the company and alter its fate. By 1985, financial services made up one-sixth of GE's annual profits — a marked jump from when Welch first took the reins and financial services accounted for about 7 percent of GE's profits, the same level it had stayed at for much of the 1970s. As early as the mid-1980s, GE's lending operation had become as large as some of the largest US financial services firms.

At its height, GE Capital produced more than half of GE's total profits. The most famous industrial company in America had essentially become one of its largest and most inscrutable banks.

To Welch and his acolytes, the proof of his methods' glorious success was in the numbers. But the glory came at great human cost. Welch famously slashed jobs wherever possible, which created tension in a company where many workers had assumed for years that they would be GE employees for life. He slashed more than 100,000 jobs in the 1980s, one-fourth of the entire workforce of General Electric, and he moved tens of thousands of other jobs overseas, where there were no unions and labor was cheap. Critics questioned whether Welch had any other management strategy than slashing costs and worried about the fallout from the dismal employee morale.

Unions and other opponents started calling him "Neutron Jack" because he removed the people while the buildings were left standing. Welch hated the nickname, but over the years, despite his fame as a CEO, that nickname has stuck to him.

Another famous and controversial tactic — often called "rank-and-

yank" — forced managers to come up with an annual ranking of the performance of their workers. The bottom 10 percent would be put on notice, and if they didn't improve, they were fired. The constant pressure from this kind of tactic only added to employee tension.

Rank-and-yank worked well for GE's acquisitions, providing a formula for trimming fat and squeezing profits out of the operations. But some managers didn't see it as helpful, especially after it had been used for a few years and some competent employees were ending up in the bottom 10 percent. You can trim fat only for so long. Also, some thought that the policy made workers fight each other for survival and inhibited managers' ability to bring their workers together to operate as a team for the good of the company. One manager tried to subvert the system by putting an employee who'd recently died in the bottom 10 percent of the ranking list in order to save another employee's job.

The company depended on its workforce as a competitive advantage, Welch argued, and keeping performance in line was an unending struggle. The GE management machine had effectively institutionalized a way for managers to overcome the natural reluctance of employees to improve on poor performance.

All of this made Wall Street happy. GE's stock soared under Welch right from the beginning and would split five times in his time as CEO. Unlike Jones before him and Immelt after, Welch was leading GE into one of the biggest business booms in history.

Some of Welch's critics see the country's economic prosperity as the major reason for his strategy's success. His supporters scoff at this, giving Welch full credit for GE's booming decade, and some even argue that GE's success actually fueled the concurrent expansion.

Workers came around to liking Welch for good reason: the stock kept rising. From 1980 to 2000, GE's earnings rose to $12.7 billion from $1.5 billion, while revenue more than quintupled, rising to $129.9 billion. The stock price rose more than forty times its value in the period.

Around him, the conglomerates of Welch's youth were collapsing, falling out of favor not just with investors but with the broader movements in American culture. The old behemoths of the 1960s and '70s were seen as unwieldy, even arrogant, in their composition, which was too broad, too slow, and too big.

GE managed to thrive in a world where its peers couldn't survive.

Giant conglomerates, like GE's historical rival Westinghouse Corporation, were fading away, and others, like AT&T, were under siege by increased competition or government regulation. GE's success only reinforced its dogma: it was the exception to the rule because of its people, culture, and tradition. The idea of breaking up GE in the Welch years was mostly seen as absurd. As long as GE could continue its long trend of hitting earnings targets and beating expectations, the company and Welch would operate with an unofficial exemption from the law of the marketplace: the right to be the last of the great old conglomerates.

Welch's returns also helped investors look past the Welch-era entanglements that didn't fit with GE's self-congratulatory narratives. The negative revelations began in the 1980s. Those black eyes included GE's 1985 guilty plea to the charge of submitting time cards for too much overtime on government contracts and a 1994 bond-trading scandal at Kidder Peabody & Company, the investment banking firm GE had acquired to boost its financial business.

Under Welch, the company was also beginning to confront ugly legacies of the past, especially the pollution it had left behind over the decades at its factories. Years of environmental activism had finally yielded laws and regulations that would require major polluters like GE to begin figuring out how to carry out massive environmental restoration projects — and preparing to pay for them.

Welch fought this work bitterly, saving a special measure of scorn for the US Environmental Protection Agency (EPA), which was pressing GE to help remove more than a million pounds of polychlorinated biphenyls, or PCBs, that GE had dumped into New York's Hudson River over the years. The dumping had once been legal, but the chemicals were banned in 1977 and were considered likely to cause cancer by the time Welch took over as CEO. Still, under Welch, the company fought hard to avoid having to pay to dredge the chemicals from the river bottom, whipping up public opposition in the small factory towns along the Hudson where the dredging would occur. Welch was livid, and he challenged the notion that the chemicals were even harmful. Either way, he insisted, the rushing current of the Hudson was strong enough that the river would clean itself.

Welch became a brand of his own as Wall Street, and to some extent even Main Street, swooned before his success and the style that brought

it. He became famous for his straight talk and confrontational approach while encouraging others to bring the same to him. As his career progressed and GE's stellar performance continued, it became harder to challenge his burning will.

Jack Welch was a bona-fide celebrity. He was treated like a star at shareholder meetings, where he would be asked to autograph the company's annual reports for adoring investors.

These years birthed the Jack Welch myth. The company's financial results were presented as proof of Welch's brilliance and his extraordinary will. They also served as evidence that GE had somehow transformed the day-to-day tasks of managing a business into something resembling a hard science.

By the mid-1990s, GE and Welch were constantly venerated by the media and the financial community, and substantial pushback was rare on almost any front. Critics on Wall Street were aggressively pursued as research and published reports produced scathing formal responses from GE's investor relations department. Executives from other companies made pilgrimages to GE's management training center at Crotonville to see firsthand a little of the internal corporate training for which Welch's company had become famous. Workouts and team-building exercises, merciless dissections of business pitches, beer-soaked, chauvinistic hijinks in the after-hours — all these features of the training helped create a whole generation of leaders forged in Welch's own image (the favored ones anyway).

The oversight role of the board was minimal. After all, Welch's meteoric success provided very little for the board to complain about. Being appointed to GE's board was already seen as a prestigious accomplishment and considered an honor for leaders, business titans, and other powerful people.

The board had largely followed the chairman's lead. One newcomer to the board under Welch was surprised by the CEO's command of the boardroom and the sparse debate among the group. Confused by how the meeting transpired, the new director asked a more senior colleague afterward, "What is the role of a GE board member?"

"Applause," the older director answered.

With a titanic image came ego and overreach. Welch had served nearly twenty years as CEO and was deep into the process of choosing

a fitting successor when he rashly decided to delay his retirement and stick around to try to buy Honeywell, a rival industrial company. But this deal was a step too far — Welch's motivation for the deal was preventing Honeywell's planned merger with United Technologies, which was a major GE competitor in the jet engine market.

What seemed to Welch a perfect capstone for his tenure struck European regulators as a brash and dangerous consolidation. The Europeans also proved far less susceptible to Welch's rough charm than the pundits and politicians in the United States. Welch seethed when he learned that this was an impasse he wouldn't be able to break. Despite GE's strenuous arguments, EU regulators blocked the deal.

The winds were shifting on Welch. GE's share price had soared for years, making it, for a time, the world's most valuable company. But now, in the CEO's final eighteen months, the share price fell 33 percent. Even as Welch drove a succession process that winnowed the ranks of his possible replacements, a reckoning was coming for the company.

Was the company — was Welch — really worth what they said?

Like his predecessor, Welch started the process to find the next CEO years before he intended to retire. He often said that his legacy would be determined by his successor, so the decision on his replacement was the most important of his career.

He wanted to take a different approach to the hunt. When he was striving get the top job, Welch had been forced to compete in a brutal battle with two other executives while they all worked in GE's Connecticut headquarters. The candidates were each named a vice chairman of the company, with the understanding that one of them would get the job and the others would remain at the company in their prestigious posts. (One of them, Edward Hood, stayed on as vice chairman with Welch until 1993.)

Jack knew that his top managers were prized throughout the corporate world. In his mind, his top talent was so good that they were better off running other large companies. In some ways, the public battle royal would simply market his best leaders for their next jobs, elevating their stature along with GE's and Welch's.

The company put together a list of twenty-four candidates in 1994 and whittled it down over the years. Some of those passed over would

have successful careers at other companies, including Dave Cote at Honeywell and Dave Calhoun at Nielsen, Blackstone, and Boeing.

The process involved the board, although Welch was naturally in the driver's seat. The board, a group who had been handpicked by Welch over the years and seldom offered resistance in the later years of his rule, went to great lengths to conceal the candidates from the public. The directors spent time with all the contenders, both with and without Welch. He worked hard to get the directors exposed to the candidates, setting up golf pairings and arranging events where he meticulously chose the seating himself.

The list was cut to eight by 1997 and then pared down to only three by 2000, when the company signaled who was on its short list by placing chief operating officers, a position that didn't exist in other GE businesses, under each of the candidates.

The three were Bob Nardelli, who ran the Power business; Jim McNerney, head of jet engines; and the youngest, an ebullient sales whiz who ran GE Healthcare and was widely seen as the front-runner, Jeff Immelt.

4

BIG JEFF

A DRIVEN, YOUNG JEFF Immelt was cruising along in the GE Plastics division in the late 1980s, wondering about his path to promotion, when an unlikely opportunity emerged. Over in the venerable Appliances business, the fridge compressors suddenly started blowing up.

As Japanese competition had begun to cut into GE's dominant position in the appliance market, GE managers had the idea to use a new device for its refrigerator compressors. The new part, essentially the engine under the hood of a refrigerator, used a rotary design rather than the traditional piston-driven unit.

The change would cut energy consumption and save customers money, just as the import models claimed to, while also creating more room inside the fridge itself. The move was revolutionary, but GE designers were confident. A similar design was already being used inside the company's air conditioners. The Appliances unit then skipped the usual field testing and rushed the new design into production, packing years of normal product preparation into a few months.

The new fridge was a huge success. But then the compressors started failing.

Not just here and there either. Under the right conditions, the critical component of GE's new refrigerators was giving out within months. Panicked engineers now said two-thirds of the machines they had sold would fail within a year and a half.

The Appliances managers had sold Welch on their new rotary com-

pressor design and even won the funding to build a new factory. Faced with the prospect of telling their tempestuous CEO that the new product was a disaster, the managers chose another route. They massaged the numbers. A new plot of the failure data showed that the failures weren't likely to reach critical mass until most of the fridges were out of warranty.

It wasn't much of a cover-up, and in truth, it didn't work. Discovering the problem, Welch fired the entire Appliances team. The refrigerators, he decreed, would be fixed, even if that meant a massive recall. Welch had an ambitious executive in mind to lead the charge: Jeff Immelt.

Immelt had been managing a few hundred sales and marketing people at GE Plastics. At Welch's order, Immelt relocated to Louisville, home of GE's massive Appliance Park factory complex, where he tried to convince seven thousand workers that they could swiftly fix the faulty compressors in more than a million refrigerators that were already out in people's kitchens.

Immelt led by diving in. He learned from engineers at Appliance Park how to make the compressor change himself. When the trucks with service crews rolled out of the gates to make repairs, Immelt was often riding along. Soon enough, the young hotshot executive on loan from Plastics was on his hands and knees in customers' kitchens, alongside the repair techs, helping to fix the flawed design, one balky fridge at a time.

Repairing a million brand-new refrigerators for angry customers was not a recipe for great quarterly earnings. The solution was expensive and time-consuming, and it required regular pilgrimages back to Welch to share with him the grim details of how long it was taking and how much it was costing. Still, Immelt had passed the test. He had been the right man to solve the problem.

Immelt's reward, at first, was a return ticket to his old job at Plastics. But Immelt knew that passing a crucial Welch management test put him on track for a promotion. Soon enough, it came: a promotion to a new post running the entire Americas operation for GE Plastics, a $6 billion business. Jeff Immelt was getting his shot at the big time.

Immelt was born on February 19, 1956, in Cincinnati. His mother, Donna, was a schoolteacher, and his father spent almost forty years at

GE, eventually managing a portion of the Aviation business, which is still based near Immelt's hometown.

His older brother Stephen blazed the trail for Jeff to follow. He was captain of the football team at Finneytown High School and president of the student council. Four years younger, Jeff would also captain the football team, as well as the basketball team.

Coming from blue-collar roots, their parents taught the boys to solve their own problems, to lead and not be victims, to "think big" and be comfortable in their own skins.

Jeff went to Dartmouth, where he played football and graduated in 1978 with degrees in applied mathematics and economics. He was president of his fraternity and worked in a Ford plant at home over the summers.

Like so many aspiring businessmen at the time, he went to work at consumer products giant Procter & Gamble, where he shared a cubicle with the future Microsoft CEO Steve Ballmer. Immelt would later tell stories of goofing off with Ballmer on the job. Immelt's next stop was Harvard Business School, but then, after graduating in 1982, he turned down a job on Wall Street with Morgan Stanley.

Instead of Wall Street, Immelt was intent on following in his father's footsteps and working for GE. It was a sound choice. GE ensured a prosperous career for workers who landed a job there. The salary and benefits were so good that some workers called it "Generous Electric."

Now, entering the 1990s, Immelt was on a well-trod path upward through the company, regularly changing jobs and assignments, mostly within the Plastics division. He felt that such moves put him in a position where the powerful Human Resources Department could get a sense of his strengths and weaknesses. Meanwhile, his personal and professional lives were melding, as so often happened at the company. Arriving in Dallas for a stint as a regional sales manager for Plastics, Immelt had encountered a colleague named Andrea Allen. The ensuing romance survived her departure for Chicago, and the couple eventually married. Andy would eventually quit GE to become a homemaker when the couple's only child was born.

Immelt himself was on the road. He spent twelve-hour days at work and 60 percent of his time traveling, squeezing in parental du-

ties on fiercely guarded weekends and evenings. Immelt was a salesman. Whereas Welch traveled to walk GE's factory floors, Jeff Immelt knew that his strength was being on the road to sell.

Immelt was enthusiastic and theatrical. Colleagues marveled at his backslapping charm, all football metaphors and eager smiles. He had been known to drop to one knee in mock pleading to seal a deal.

He hurled himself with equal abandon into the team-building exercises and the retreats through which GE, under Welch's tutelage, tried to forge cooperation among its far-flung executives. It was at these same events that the company usually scouted for leaders.

So there was Immelt, burly and towering over the others, in a yellow life jacket and blue bandanna on Hurricane Island, in the fog and frigid waters of Maine's Penobscot Bay. In a GE-sponsored visit to Outward Bound, teams were competing over a single long weekend, building rafts out of wood and barrels and downing green bottles of imported beer at day's end. Immelt, the brawny former lineman, was the biggest man on his team's raft, and some of his teammates credited him with naming the group as well: the eight-man band would be called El Pulpo, after the eight tentacles of an octopus.

In the middle of one sailboat race, in which the team was losing right from the start, the crew of El Pulpo tried augmenting their pace with oars. Cooped up in the long Hurricane Island boat and subjected to one too many splashes from Immelt's crabbed oar, a company labor negotiator named Dennis Rocheleau finally snapped.

"You splash me one more time, I'm going to stick that oar so far up your ass your tonsils will pop out," Rocheleau hollered at Immelt as the other occupants of the boat stared in disbelief. Immelt leapt to his feet, the boat rocking as he jutted his chin into Rocheleau's face before the two of them were separated. This was just the fiery vibe that Welch liked — an ability to clash fiercely in competition, only to relax again over beers at the end of the day. The altercation showed the traits that had put Immelt on Welch's list of rising young bulls in management who wouldn't back down from a challenge.

A few days later, Rocheleau walked away from a long climb up the face of a granite quarry, sputtering and fuming at the exercise, only to stumble upon Immelt, who had taken the long way and become lightly

stuck in a crevice. The future boss, scraped and bloody in a white track-suit, was caught between some jagged rocks. Rocheleau offered him a hand to help him out. Jeff Immelt refused.

Running a manufacturing operation was a new experience for Immelt, and he had a lot to learn. He came to the new post in the Plastics division in 1994, the year he often points to when describing his education as an executive. That's because he came close to getting fired.

What exactly happened in that time is controversial. As Immelt tells it, the division was caught in a squeeze as its raw material costs rose, crunching profits because it was locked into fixed-price contracts. Profits that were targeted to grow 20 percent rose by only 7 percent, and Immelt missed his goal by $50 million.

When GE's annual leadership meeting in Boca Raton came around in early 1995, Immelt worked hard to avoid the boss, who would inevitably confront his underling about his poor performance. Instead of boozy nights and golf-filled days, Immelt made sure to slink into dinner late and then went up to bed early. On the last night, as he snuck off to the elevator, a firm hand landed on his shoulder. There was no question that Welch had found him.

"Jeff, I'm your biggest fan, but you just had the worst year in the company," Welch told him, as both men have separately recalled. "Just the worst year. I love you, and I know you can do better. But I'm going to take you out if you can't get it fixed."

Immelt might have failed at avoiding the boss, but persuasion was his real strength. He told Welch what he wanted to hear.

"If the results aren't where they should be, you won't have to fire me," he told Welch, "because I'm going to leave on my own."

Welch's wasn't an idle threat. Dave Cote, another young executive phenom who was the same age as Immelt, had lost his shot at replacing Welch because he wasn't making his numbers when running the Appliances business.

As with most things at GE, it was more complicated than it first appeared. When Immelt took over the Plastics operation, the previous management hadn't been playing it straight. Under pressure from Welch, the division had stretched to make its numbers, including misreporting inventory figures to reduce the cost of goods sold. Fudging the

numbers about the size of the inventory gave an artificial boost to the division's earnings, making it appear to have made the numbers Welch demanded.

The problem stemmed not from any single action but from the practices of accountants on staff at the dozen or so plants in the division. They'd reported their numbers by working backward: starting with a profit target and then working out what their sales figures would have to show to get there, rather than simply running the business and reporting their results to headquarters every three months.

Immelt didn't have a chief financial officer helping him in his first few months running the Plastics unit. Unfamiliar with the intricacies of the business, it was months before he understood the inventory tricks that the business had been playing to hit its targets. When Immelt did comprehend what was happening, he faced an immediate dilemma. There was pressure from above to not make waves over the issue, and coming clean to headquarters about Plastics sending them faulty financial results could dent Immelt's image as a gifted manager, just as he was trying to establish his presence as a rising force in the corporation.

Immelt figured that if he had managed to get a million refrigerators repaired, he certainly could work his way out of this fix. So he swallowed the problem himself. It wasn't just the flawed accounting that Immelt would have to find a way to quietly set right. For a GE executive, few things mattered more than financial projections — the size of the increase in sales and profits that the executive would commit to making for the company in the next quarter, or over the next year. Because his division had been gaming numbers to make their results look better than they were, Immelt was going to have to come up with two years' worth of strong earnings in half that time just to keep pace with the targets that Plastics was already projected to hit.

The stress was overwhelming for Immelt as he dealt with this demonstration of how Welch's numbers-obsessed culture could go wrong. Welch would argue that he pushed his underlings to produce results, not fraud. But even if the CEO didn't bend the rules himself, Welch cultivated an environment of pressure that incentivized people to do just that. And his rough discipline also discouraged executives like Immelt from coming clean about their actual results once they recognized that

they were in deep trouble. To do so might have been the more honest approach, but it also almost certainly would have blunted, if not ended, an ascendant career. If you couldn't do the job and hit your targets, they all knew, Jack would get someone else who could.

Rather than clean up the books of the previous management, Immelt pressed ahead under the old projections — and promptly missed his numbers. It was a black mark, but ultimately just a hiccup, not the career-altering detour it could have been had Immelt reported that the numbers in Plastics had been fudged for years. Instead, the explanation was mundane: rising costs of raw materials had reduced GE's profits on contracts whose prices were fixed. Immelt lived to fight on.

The flip side of Immelt's jock-like salesman bravado was his aggressiveness, which almost triggered another disaster. Trying to boost the numbers from his division, Immelt pushed General Motors hard for a price increase on the plastic materials that GE supplied for GM cars. When Immelt and a GM executive nearly came to blows over the price dispute during a meeting at a restaurant, the young executive's desire for standout numbers almost destroyed GE's relationship with a vitally important customer. The relationship held, however, and Immelt, as was his way, laughed off the incident. The path he took had been messy, but the result was right. He told everyone who would listen that his priorities had been fully realigned by his time at Plastics. Immelt would never again overlook the minor details that went into making his company's numbers.

Welch was forgiving of mistakes and knew what it was like to be in the hot seat. Early in his career in the same Plastics division in Pittsfield, Massachusetts, Welch was in charge of setting up the large-scale production of a new product. While a pilot facility was being set up, the main tank exploded and literally blew a hole in the roof.

When he was called to headquarters to explain himself, Welch was figuring that he might be fired, so he was surprised when his boss expressed more interest in learning about what went wrong and how they could fix it. Welch would say that he learned from this episode that it accomplishes nothing to step on the neck of someone who is already down. Despite his reputation for ruthlessness, Welch wasn't heartless; he understood that a good coach knows when to keep his mouth shut and let the lesson be taught on the field.

And out on the field, Immelt was learning plenty. The young executive was now running the GE Plastics operations in North and South America. In the Americas, the company had factories on the ground that made all the basic materials it needed to manufacture its products, which were mostly plastic pellets for molding into different forms. Immelt's counterpart in Europe did the same, relying on GE factories in France and the United Kingdom, but the Asian operation didn't produce what it needed and instead purchased supplies from the other regions. In the organizational wrinkles of such arrangements, opportunities for ambitious salesmen were born.

Similarly, a transfer of GE Plastics components from the Americas division to Asia didn't bring any actual money into the company. But accounting was accounting, and on the company's ledgers those transfers were registered as sales at the regional level. Near the end of some fiscal quarters, the phone on the desk of another talented young GE executive, John Rice, who headed the Asia-Pacific division in Singapore, would sometimes ring. Without fail, on the other end was Jeff Immelt, asking for a favor. Just a few internal orders would be enough to help Immelt beat his earnings targets and keep him on the ascent toward an even bigger position.

5

THE RACE

A S THE SUCCESSION race for Welch's job heated up, the finalists were put into new positions to audition. To present him with another test of his ability to run a major business, Immelt was named CEO of GE Medical Systems. The business was a core part of the conglomerate and a market leader in CT scanners, MRI machines, and other medical equipment and supplies. For Jeff Immelt, it meant a move to Waukesha, Wisconsin, just a few miles west of Milwaukee. There he would manage twenty thousand workers — four times his workforce at Plastics.

To some, Immelt was the shoo-in for the job. He had the edge that Welch loved so much, exuding self-confidence in his ability to complete any task. This was the swagger that Welch cherished. Immelt carried himself like a winner.

Immelt flexed his sales and marketing muscles in the new job, but he also went deeper. He plowed money into developing new products, always with the goal of expanding GE's capacity to sell service contracts that would keep the equipment running for years. Selling massive machines to hospitals and clinics was the mission, but the real money came from long-term service, which provided a steady stream of predictable revenue and also kept the sales team engaged with their customers. For a salesperson, just getting time in front of a customer can be a major challenge. When customers know you by your first name, the company's

thinking went, it becomes a lot harder for them to start buying from the competition.

Beyond providing repairs and maintenance, Immelt pushed the division to work with customers to help improve productivity. This kind of attention helped shift the customers' overall impression of the company and made them feel like partners with GE, which would be ready to do business at any time. Immelt also pushed the division to expand overseas.

Meanwhile, Jeff Immelt, like so many other business executives in the 1990s, was falling in love with the internet and trying to figure out how to get some of the vibrant energy of that sector — along with the rapturous coverage in the business press — for his own business and for himself.

As Immelt and his peers were rising up through the hierarchy of GE, being graded on the conventional marks of quarterly sales and earnings, a revolution was rippling through the rest of the American economy. New companies like eBay and Amazon were mushrooming into existence seemingly overnight. They lacked the long-term predictability and stability that was the hallmark of blue-chip corporations like GE, IBM, and DuPont. Unlike these established companies, it wasn't clear how the new tech firms were going to make money decades hence. But it was these firms, riding the popular wave of internet connectivity and software innovation, that had all the momentum, verve, and aura that GE lacked.

All around them, companies with only the loosest tangential connection to the internet were dropping "dot-coms" and "webs" into their promotional materials. They hoped that using the buzzwords of the late 1990s would spread some of the internet excitement to their own businesses. And sometimes it worked. Sometimes just being associated with certain internet trends could make a company's shares soar, seemingly with no regard to its business fundamentals.

Jack Welch was not impressed. He scoffed at the notion that the new dot-coms could run with a company like GE. Startups like Amazon were building warehouses, distribution systems, and manufacturing facilities — all aspects of doing business that GE had long ago perfected and knew how to do while preserving the all-important profits. It seemed to

Welch and others of the old school that the rules of economic gravity would eventually kick in for these internet companies, and they would find that they were just another competitor in a crowded marketplace that GE believed it dominated.

That didn't mean that GE didn't learn to talk about the internet just like everyone else was. For a marketing-oriented leader like Immelt, it made sense to play up any connection GE had to the subject that had captured the world's imagination. And Welch also urged GE business leaders to not fall behind, pressuring them to make sure their divisions had websites and an internet strategy.

GE's annual report mentioned the word "internet" just eight times in 1998. One year later, the word appeared fifty-six times. But then the buzzword would disappear just as quickly as the '90s internet stock market bubble did. The word "internet" was back down to ten appearances in the 2001 annual report, having rippled through the company's description of itself to the share-owning public and then disappeared again.

But the internet craze had a more lasting effect on the company in another way. It made the youngest, most outwardly charismatic candidate to succeed Jack Welch that much more appealing. In a company run mostly by gray-haired white men, Jeff Immelt's relative youth seemed a more essential quality for the leader who would guide GE in a new business world, one where the leaders of some of the hottest tech companies still styled themselves and spoke like undergraduates. Immelt was just forty years old when he was handed the Medical Systems business, which put him close to the same age Welch had been when he was named CEO of the entire company.

It wasn't just age working for Immelt. His success also helped. In less than three years as CEO of GE Healthcare Systems, the division's revenue shot up to $6 billion from $4.2 billion, fueled in part by the smaller companies that Immelt acquired.

To the doubters — and his rivals for the job — Immelt was untested and probably unready. His experience was anchored in marketing and sales, and mostly in the Plastics division. Plastics had once been a large part of GE, which invented products such as Lexan, a hard plastic, but that division wasn't seen as the future of the company.

Immelt had no experience in the crucial Power division or in his

father's former home in the company's Aviation business. GE Capital workers were concerned that Immelt hadn't spent time working with them, even though they brought in 40 percent of the total GE profit. And he hadn't worked at GE headquarters since the early 1980s. Immelt's detractors thought it made sense to have a more seasoned company man follow Welch, one a little bit older, a little less ambitious and eager — someone a little less like Welch himself twenty years earlier.

In theory, the succession process at GE was a secret, and the company did keep its cards as close to the vest as it could. But Welch was a global celebrity at the tail end of his tenure, and the race to succeed him was being closely watched by the press and the public. As always, Welch was soaking up more than his share of attention. And so it was clear by 1999 who the three men on the short list were: Bob Nardelli, who was running the power turbine business; Jim McNerney, the laconic head of the jet engine unit; and the young man who came up through Plastics and had run Medical Systems for three years, Jeff Immelt.

Like Immelt, McNerney came from an Ivy League background and had degrees from Harvard and Yale, where he had been a pitcher on the baseball team and a frat brother of former president George W. Bush. The product of a large family — McNerney's father had been CEO of Blue Cross/Blue Shield — he kept a lower profile than his rivals but was a cool operator. He spent his time at Aviation cultivating a better relationship with Boeing but also spoke with other companies about possible leadership positions, just in case things didn't work out at GE.

Nardelli, the most outspoken of the three, was a strong performer who always made his numbers. He came from solid blue-collar roots — like Immelt's, Nardelli's father was a GE veteran — and the younger Nardelli had long dreamed of being CEO.

Nardelli campaigned hard for the job. At one point in 1999, Welch and other executives were visiting Schenectady, New York, the home of GE Power. Riding into town, the group was startled to pass a billboard with a giant image of Nardelli's face looking down at them. He had personally commissioned the ad to commemorate GE's support of a local charity, but the image also made it clear that the Power chief was running for something.

The board would take the final vote on the next CEO, but ultimately the decision was up to Welch. And despite more than five years of rig-

orous planning, evaluating, and analysis, Welch couldn't quite explain why he chose as he did. Maybe age was the factor, company insiders whispered. Like Welch when he took the job, a younger chief executive would have a long expanse of time in which to grow into the job, learn where to pilot the company, and see long-term bets pay off in commercial success. Or perhaps he simply preferred a candidate in his own image — an unorthodox choice with rough edges, rather than a carefully refined product of the company's rearing. But Welch was Welch, so in the end he went with his gut. His choice was Immelt.

The board had seen one warning sign about Immelt's qualifications and performance: he often paid too much for acquisitions. Immelt tended to start negotiations too high, sometimes to the surprise of others involved in the deal, leaving little room for negotiation. It wasn't uncommon for the board to approve one of Immelt's deals, only to have him ask for approval to pay more in order to make the deal work. In some ways, this tendency simply reflected Immelt's experience as a salesman. He'd always needed to close deals, and for a company like GE, paying a little more didn't seem to cause any concern.

A *New York Times* profile of Immelt in 2000 described his refusal to lose deals. Once he flew to a hospital whose executives had decided to use competitor machines to try to persuade them to change their minds. In winning the deal in the end, Immelt "did not give an inch on price, but threw in some extra services and features."

This was the traditional way to close a deal — the car salesman throwing in the floor mats. But those floor mats cost something, and the addition of extra services and features to the deal also had a cost for GE.

The decision was officially made on Thanksgiving weekend of 2000, while the rest of the country was focused on a still-unsettled presidential election. The board's support for Immelt was unanimous.

There is no science to picking the CEO of a company. Predicting a single person's success at juggling uncertain events in dynamic markets spread across world economies can only be done with a time machine. At large companies, the succession decision is made by the board, which nominates its own members. At GE, the CEO also serves as chairman, giving that person even more influence over decisions. In such a corporate culture, it isn't difficult to see how bad decisions can compound or flawed strategies can become entrenched.

Many of the board members and advisers from the time of Welch's retirement stand by the process. Even those who were loudly critical of Immelt conceded that their choice still seemed correct in light of the information they had at the time.

When the moment finally arrived, Welch delivered the news to Immelt and had him down to the Welch house in Palm Beach to celebrate with his family along with a few other top executives. The secrecy was so tight that they didn't dare go out to a restaurant. They even used a contract plane rather than the usual GE executive jets, for fear of leaks. Immelt was listed on the flight manifest as the son of a longtime director.

Later that weekend, Welch made a whirlwind plane trip to drop the bad news on the two runners-up. In his memoir, he details the emotional difficulty of having those conversations. In a private airplane hangar on a rainy and foggy night in Cincinnati, he told McNerney, who took it well and complimented Welch on a well-run process. Later McNerney would say it was the most disappointing moment of his career.

Nardelli learned his fate in a similar way, at a private airport near Schenectady, New York, but he didn't take it as well as McNerney. He immediately pushed back, demanding to know why he had been passed over. Nardelli argued that he had met or exceeded every target he could imagine. He also felt that he had the operational chops that Immelt lacked.

The disappointment of finishing in second place amplified the bitter taste that Welch's tournament had stirred up all along. The danger of engaging in such a fierce and long succession contest, both publicly and privately, was that the candidates would put all of their efforts into the goal of becoming CEO, leaving little option for the vanquished but to fall on their swords in defeat. Like a presidential election, there really was no second place.

This wasn't an election, however: General Electric was a company that needed to have leaders and their institutional knowledge sticking around. Instead, the former contenders had plenty of external options to choose from after Immelt was chosen. Welch had even worked behind the scenes to make sure that his boys would be sought after, talking them up to fellow CEOs. Those other companies now jumped at the chance to poach the progeny of Welch. Within days of being passed over to succeed their boss, Bob Nardelli left the GE he'd once dreamed

of running one day and signed up to serve as the CEO of Home Depot. McNerney left as well, to lead another industrial conglomerate, 3M.

In Jack Welch's GE, a business wasn't worth being in at all unless you could be the best or the second-best. But among the generation of leaders being groomed within his company, it turned out that even being second in command wasn't good enough.

At the press conference after the announcement in November 2000, Immelt and Welch wore matching dark blazers and blue shirts without ties. They claim it wasn't planned. The six-foot-four Immelt, with his graying plume of hair, towered over Welch.

With them were two vice chairmen, GE Capital's Dennis Dammerman and NBC's Bob Wright, their presence sending the message that the baby-faced forty-four-year-old onstage wouldn't be alone as he inherited one of the biggest and most complex companies of our time.

Immelt was presumptive chairman, just the tenth in the history of the company, but the job wasn't fully his yet. Welch had his successor wait awkwardly in the wings for months, having delayed his retirement to try to close one last big deal.

While Immelt eased into the title of President and Chairman-Elect, Welch tried to hammer home the massive acquisition of Honeywell, an industrial counterpart that would vastly expand GE's position in the aviation market. For ten months, Immelt pitched in, taking to the airwaves to tout the benefits of the Honeywell deal or to play down persistent rumors that NBC was going on the block. A media blitz was playing to Immelt's strengths. He was making connections and spreading his wings. The incoming boss met with customers, gave facetime to Wall Street analysts, showed up in press release quotes, and joined conference calls.

And then Welch did something unexpected. He failed to make the deal. Casual observers of the company had grown accustomed to assuming that Welch would always find a way to get what he wanted, even a gigantic late-career acquisition like Honeywell that he wouldn't be around to see through to the final stages of assimilation — the merging of businesses and workforces, the layoffs and restructuring, the victorious dropping of new revenues into profits on the GE bottom line. But Jack Welch had deeply misunderstood the degree of resistance to a massive American aviation merger in Europe, and the painful severity of the concessions that European Union regulators would demand in order to

let the deal go through. The deal had been such a foregone conclusion that some departments at GE were already working with their counterparts at Honeywell, where some of the workers were reporting to new bosses as if the deal had already closed. But Welch, to his shock, soon realized that the European concession demands had pushed the deal beyond the point of reason. At a certain point, even a deal he coveted as badly as Honeywell didn't make business sense, no matter how much they had touted it to Wall Street.

Welch took his cue. It would be another month before the companies formally called off the deal, blocked by the European Union's resistance. But the most famous CEO in GE's history wouldn't be waiting around for that announcement. Jack Welch left the office for the last time on September 7, 2001.

Jeff Immelt alone was now in charge of General Electric.

6

MAKE SOME NOISE

O N JEFF IMMELT'S first day running the company, GE workers streamed into the concrete-and-steel monolith of the Bradley Center in downtown Milwaukee, the aging home court of the NBA's Milwaukee Bucks. The cavernous space dwarfed the crowd of four thousand workers who arrived by bus to hear their new CEO speak.

Most of the workers came from GE's medical equipment business, GE Healthcare Systems; Immelt had held his most recent job at its headquarters just a few miles away before being crowned to succeed Jack Welch. The forty-five-year-old new boss wouldn't actually be in the building that Monday; instead, he would address the crowd via a massive video screen hookup, beamed in from GE's management training complex back in Crotonville, New York, as cameras in Milwaukee relayed the Bradley Center scene back to him.

In addition, one-third of GE's workforce of 300,000 employees around the world watched as Immelt addressed them for the first time. In an open-collar shirt and sweater vest, Immelt spoke for two hours, hitting all his bullet points about GE moving fast in the modern world, embracing the internet, and, of course, winning.

For its new boss with a jock style, GE held a pep rally, an unlikely spectacle of a throng of corporate workers cheering for upper management. No detail was left up to chance. Just before Immelt went on, two young women hurried into the arena in Milwaukee to stir up the crowd.

Like stage crew members at a TV talk-show taping, they held up signs to the crowd of GE workers that read: NOISE.

After four decades at General Electric, half of them at the top, Jack Welch retired a celebrity. In his first week after handing over the reins, he was in New York, where he had a sprawling apartment on the forty-seventh floor of the Trump International Hotel and Tower, a building GE itself had owned and redeveloped under the Trump banner. On the corner of Central Park, the lavish home had breathtaking views of that ocean of green in the heart of Manhattan.

The previous Friday, GE's board had convened with Welch as the chairman for the last time, and in his last official task, Welch had formally appointed his successor, Jeff Immelt. A retirement party followed at Crotonville, the cultural center of Welch's reinvention of the company.

On Tuesday morning, September 11, the sky was crystal clear blue over New York. For Welch, it was a day long in the making. His book, *Jack: Straight from the Gut,* was going on sale as perhaps the most anticipated business book of all time, and Welch was kicking off the launch with an appearance on *The Today Show.*

Since Welch had put the show's host, Matt Lauer, in his job — a tale he included in the book he was hawking — he was assured a warm welcome. No one would be asking hard questions about the Honeywell deal. The book was a hit before he even wrote it. Just about everyone wanted to hear the inside story of GE from the mastermind who built it.

Welch's track record was legendary, and he was surrounded by people who lavished on him as much praise as possible. Despite the Honeywell trouble and a few months of downward drift in GE's stock price, Welch was celebrated for his successes and declared a saint in the business world. The New York Stock Exchange held a dinner and cocktail party in honor of his career. To Welch's devotees, the company's triumphant decade was proof of the efficacy of the GE management machine he had assembled. Welch, in this telling, was the ultimate leader — a model for the next generation to emulate.

The company's soaring valuation through the 1980s and '90s had made high school–educated factory workers into millionaires, Welch said. GE had long encouraged employees to ride along with the soar-

ing trajectory of its stock, offering savings plans that accumulated GE shares at a discount or with subsidies as GE's stock outperformed virtually all others.

The company's mystique had been amplified in the figure of Welch himself. Adored by the press and Wall Street, Welch was a household name, with a story to tell about positioning a century-old industrial company for the dawning of the information age. He would be right at home on a national morning television show.

When the first airliner hit the World Trade Center, Welch was at 2 Rockefeller Plaza, just across the street from the GE Building, waiting to go live on *The Today Show*. Welch never got on the air.

As TV channels around the world dialed in live feeds of the attacks on New York, Washington, and Pennsylvania and cell-phone networks in Manhattan creaked and then gave out under the deluge of calls, Jack Welch, the most famous boss in America, wandered out onto the corner of Forty-Ninth Street. He watched from a vantage he shared with hundreds of thousands of New Yorkers who, alone or in groups, stared south down the stilled avenues, in tears and stunned silence, at the giant plumes of smoke and flame. Welch was as cut off from information as the people around him on the Sixth Avenue sidewalks. Cable chyrons blazed rumors and best estimates, but few had any idea how many attacks were occurring or still might. The roar of military aircraft shook the blue skies over New York.

That moment, Welch thought, made the launch of his book seem "like the dumbest, most insignificant thing."

Day broke over Seattle on Jeff Immelt's first Tuesday morning as GE's CEO. The salesman was on a sales call in the Pacific Northwest to meet with Boeing, the world's largest plane maker and one of GE's most important customers.

Immelt was in the Pacific Time Zone in the midst of his morning routine. On this day, that was exercising on a StairMaster while devouring live TV coverage of the inferno in the North Tower when the second plane struck. As a consensus immediately registered among cable news reporters that an accident was an impossibility, it dawned on Immelt that, with air travel indefinitely halted, he was stranded in Seattle,

with other GE executives scattered around the country, on his fourth day on the job.

The new CEO was quickly on the phone to heads of the business units and the board of directors. He also called Jack Welch, eager for advice. Immelt knew that, faced at the beginning of his watch with a global catastrophe that would have huge effects on every level of the company, and before the markets and the public had had a chance to take his measure, how he responded would be formative. It would also be a test for the fortress of GE, which had found prosperity through wartime and peace.

From the first moments on the StairMaster when he saw the news, it was clear to Immelt that the attacks would hit General Electric hard. Its insurance business was one of the companies covering the destroyed buildings, its TV business would feel the sting of lost ad revenue and absorb a surge in reporting costs, and the indefinite suspension of air travel would strangle the sales of GE spare parts and repairs of its jet engines, which now sat idle across the country.

But the real hit would be deeper. The economy was already heading into its first recession in a decade, putting undeniable pressure on GE, and the attacks would only exacerbate that downturn.

Meanwhile, even Jack Welch's stock was falling. With his era as CEO concluded, Welch now stewed as his messy divorce played out in newspapers and legal filings, revealing embarrassing details about lavish perks and spending — fresh-cut flowers flown in from Europe, divorce by fax — that reflected badly not just on one rich couple but on the company that had permitted and often paid for it all. As in his "Neutron Jack" days, this was another easily grasped portrayal of what Welch looked like up close that was not very flattering. Even more galling were the questions, as the stock price swooned from its peak under Welch, about whether the very best of the golden years hadn't been driven by creative accounting rather than his genius. These doubts were quietly voiced at first, but something had to explain an otherwise incomprehensible streak of earnings growth. Now the obligation to explain away such concerns sat squarely in the lap of Jeff Immelt.

Like his predecessor, Immelt had shown an ability to shake off criti-

cisms and unwelcome questions. And nothing whisked them away more quickly than financial success. Now, despite his youth, he had to keep up the same impregnable defense that Welch had managed and navigate the changed state of a company whose detractors wondered where it still fit in a shifting global economy.

Immelt quickly adjusted to being in the club of prominent CEOs, the handful of (mostly) men who ran the world's biggest enterprises. He devoured books and consumed a steady diet of news — heavy on business and sports. Like Welch, he relied on humor and an emphatic plainspokenness to maintain support among the rank and file. And Immelt still clung to his Ohio roots. His language was peppered with the jargon of the Finneytown High jock he once had been as he enthused about "blocking and tackling" from the stages on which he held forth to GE investors.

It was a belief in the wisdom of the playbook that had informed Immelt's every step along the path to becoming CEO. Even in his early days, peers felt, Immelt had seemed so confident in GE's operational excellence — the quality it claimed it could instill in any manager and teach to customers and colleagues alike — that he hadn't always seemed to burrow into the details of the particular industrial unit he led. He'd been able to manage through crises and goad his team to success using only GE's management machine and his own positive and aggressive browbeating to stretch goals and hit higher targets. It was the GE way.

Structured as a series of silos in a matrix management scheme, GE was a corporation with multiple, complex, and entangled management hierarchies. Workers generally reported to more than one boss. For example, chief financial officers of a business division reported to the leader of that division, but also to another supervisor in GE's Finance Department, which was led by the CFO of the corporation.

Immelt had risen in the company from the ranks of those who sold GE's products and services. He had only rarely been responsible for the intricacies of production problems, operational inefficiencies, research, or finance. Nevertheless, like Welch, Immelt had no problem trusting his gut. But it could be a weakness. When deciding on a deal or a move into a new line of business, his peers said, he rarely strayed from his initial position, and being presented with contrary evidence by internal

analysts often visibly annoyed him. Immelt had a tendency to blow off bad news.

"That's just your opinion," he would say. "You're not looking at it right."

But times were now changing, and changing rapidly — in almost every conceivable way. The terrorist attacks marked the dawn of a new era of uncertainty and fear in the United States. There was no question that the period of prosperity and safety in the 1990s was over, and even the certainty and authority of General Electric would be subject to the storm of swirling forces.

Stranded in Seattle, Immelt was working numbers with his advisers. A global panic had grounded thousands of planes that GE rented to airlines, along with tens of thousands of GE engines, which spit out profits only if they were in the air. Its insurance business was wavering, and the prospect of global instability dampened the prospects for the major investments that buoyed GE's power and medical device businesses.

A few days later, the GE investor relations shop emerged with a number for the immediate cost to General Electric, financially, from the September 11 attacks: $600 million, meaning the next quarter's earnings would be only slightly less than what analysts were expecting.

Like most of corporate America, GE was hit hard — it also lost two employees in the attacks — but it did find a marketing victory as downtown Manhattan was still smoking. Chief of communications Beth Comstock, a forty-one-year-old public relations climber who came to GE from a similar role in NBC's news division, had been called up to the big leagues by Jack Welch, who was looking for someone to run communications as he prepared to hand off the baton. Immelt inherited Comstock, and she clicked right into place with Immelt. In those first days, she helped develop an advertisement to declare GE's patriotism and show support for rebuilding after the attack. It was a sketch of the Statue of Liberty rolling up her sleeves and getting to work.

Immelt liked it and told her to go ahead with it. But BBDO Worldwide, GE's longtime ad agency, and pretty much everyone else at the company who saw the idea hated it. Those opinions didn't matter.

"We listened to everyone. But in the end, we overruled them," Com-

stock wrote in her memoir. The ad was a big hit. Its success over the chorus of objections was a major victory for Immelt, who appreciated leaders who went with their gut and stuck to their guns. The success of the ad also set the tone for Comstock's trajectory at the company.

Immelt had been handed the keys to a house that immediately caught fire. It was his company now, the new CEO felt, and the new assaults felt personal, and as if they came from all sides.

"My second day as chairman, a plane I lease, flying with engines I built, crashed into a building that I insure, and it was covered with a network I own," he said just weeks afterward.

Immelt wasn't just watching from the sidelines. The attacks instantly flattened the airline industry; not only was the four-day halt to all air traffic crippling, but passenger reluctance to fly would last for years. GE teams helped carriers adjust to the downturn and provided crucial financial support by allowing customers to defer payments and advancing $5 billion in direct financing to the airlines.

In the aftermath, Immelt was learning what Welch had known: how much GE's CEO could lean on GE Capital when times got tough. There was an option in this fraught moment as stock markets warily reopened and companies lumbered back into operation, trying to locate a new normalcy: take a giant charge. A massive loss in the accounting would reflect all the financial loss that GE knew about in the wake of the disaster — its engines, its insurance book, its macroeconomic outlook. It also might reflect the lingering uncertainty that GE's investors did not fully understand, like the weaknesses in its opaque financial arm that couldn't be valued by everyday investors, and its knowledge that it probably had never quite been worth what it seemed to be at the high-water mark. A massive loss might trigger a sell-off in the company's shares, but it would also serve as something like a reset — an understandable adjustment in a moment of national reckoning, and perhaps a new beginning.

But GE opted to soften the blow. Only later would it be obvious that 9/11 did more harm to GE than the company acknowledged at the time.

The company's fourth-quarter earnings in 2001 managed to rise nearly 10 percent, hitting revised Wall Street expectations. The jet engine business was hit hard, and so was NBC, but the strong financial services business helped to offset that weakness at just the right time.

Immelt went on *Fox News* in January to declare victory, citing GE's

strategy of owning diverse businesses to offset tough times. "I think it just works," he said, pointing to the company's digitization and well-known Six Sigma practices. "And we have just great financial discipline to allow us to do acquisitions and continue to grow in down cycles."

But GE's fortress of diverse businesses couldn't ward off the economic downturn and the corresponding drop in demand for GE's sweeping portfolio of products. Layoffs were coming.

THE EDISON CONDUIT

IN EARLY 2001, GE Capital was buying up real estate assets in varied markets, spending billions per deal to scoop up operations in every corner of the country. These were some of the last deals under Jack Welch, and they injected GE into the world of lending money to restaurant chains like Burger King and Cracker Barrel, as well as businesses like Midas Muffler Shops and the Circle K convenience stores.

The acquisitions brought in thousands of loans connected to properties sprinkled across the land and were a far cry from making jet engines and selling refrigerators. It was just the kind of business needed to feed the fire of GE Capital, which was now one of the world's largest nonbank finance operations.

A year later, Moody's put $164 million of GE's asset-backed securities on watch for downgrade, citing "a rising percentage of problem loans in the pools" behind those instruments. But it was size that mattered, and GE wasn't slowing down in its plunge into real estate.

The push into real estate didn't stop with Immelt taking the reins. In December 2001, GE paid $4 billion for Security Capital Group, adding to its real estate holdings in GE Capital. It also formed a joint venture with Kimco Realty Trust to buy up neighborhood shopping centers. By the end of that year, GE's commercial real estate group had grown to about $24 billion.

Inside GE's legendary management machine was a complex mechanism that used these deals to help the company meet its profit goals.

Cash flow wasn't as important as the all-important earnings per share results that made Wall Street swoon for GE every quarter. In the summer of 2001, Immelt had even declared on CNBC that GE wouldn't disappoint Wall Street in the near future. "There is not going to be [an earnings] miss," he said on the air.

He spoke with confidence for a reason: GE Capital had his back. Managing financial results wasn't unique to GE, but the degree of GE's reliance on the practice was. Management, with its customary swagger, treated the frenzy of last-minute tweaks and transactions each quarter as entirely natural.

This way of managing financial results wasn't Immelt's invention, and it wasn't a secret. GE executives have acknowledged that they worked to make sure earnings were always growing in a nice smooth trajectory. "We think consistency of earnings and no surprises is very important for us," former CFO Dennis Dammerman, one of Welch's most essential lieutenants, told *Fortune* magazine.

A rotund Iowan who joined GE in 1967, Dammerman became CFO in 1984 and oversaw much of the financial services growth. For some, Dammerman was the safety net that kept the financial side of GE from spinning out of control. He collected classic cars and racehorses and used GE's fleet of corporate jets more than any other executive. He was "aggressive in every sense of the word," said those who worked with him, and he also kept the rank and file in line. Upon hearing a proposal for a project, he might bark back, "That is the dumbest fucking idea I've ever heard." The shocked recipient would then be forced to defend their reasoning, and Dammerman could soften his stance upon hearing a convincing argument. But Dammerman's rough style discouraged new ideas that were not thoroughly vetted.

Dammerman defended GE's opaque structure and its curiously solid profits with a dismissive confidence that bordered on hubris. Only GE could understand how its businesses made money, and investors should simply be happy with the results.

"We're a very complex, diverse company that no one from the outside looking in can reasonably be expected to understand in complete detail," he said. "Our story to the investing world is, we have a lot of diverse businesses, and when you put them all together they produce consistent, reliable earnings growth."

To achieve this, performance targets were set first, and only then was the method for achieving them determined. The contributions to making the numbers were spread around the various businesses, and compliance was required.

A steady stream of acquisitions fed the earnings momentum. GE could use its unusually high price-to-earnings ratio for an industrial company as high-value currency to pay for deals. By acquiring companies with a lower price-to-earnings ratio, GE was getting an automatic earnings boost.

As an example, if GE was trading at a price-to-earnings ratio of 40, that meant that, if its stock was $40, it was earning $1 per share every year. If GE then bought a company with a price-to-earnings ratio of 10 — that company was earning $4 per share for every $40 of stock — GE was essentially trading $1 of earnings for $3 of new earnings without doing anything except making the deal.

Using cheap financing can perform the same trick. GE could buy a company that produced profits at a rate exceeding GE's cost to borrow, which was very low.

GE shares weren't going to stay valued that way forever. It would have to do more deals or use other methods to produce earnings. One way to do this was by contorting accounting rules to make acquisitions seem even more profitable.

For example, it could buy a real estate company with thousands of properties and appraise them at a low value, a strategy that reduced the earnings hit from depreciation and brought a huge profit if the property was sold at a higher price. But that method works only if the buyer is willing to pay. GE Capital had a solution.

Meanwhile, one of the biggest companies in America was imploding. Enron, a Texas pipeline company that had moved into energy trading and other ventures, including broadband, was a self-proclaimed symbol of corporate ingenuity at its peak, bringing in more than $100 billion in revenue. But now Enron had been brought down, seemingly overnight, in a massive accounting fraud.

Enron executives were self-dealing, earnings were manipulated, and its books were a mess. Complex structures had been used to hide debt, and valuations of assets were pure fantasy. The company would arbitrarily decide when an investment was going to produce earnings in

the future, then calculate its current value. Enron's stock collapsed as the scheme became apparent, sending investors and other corporations running for cover.

The corporate accounting world had looked very different before the fall of Enron. Enron had seemed unbreakable, with direct connections to the White House, and enjoyed widespread approval for using what it said were innovative strategies to supercharge its profits in the internet age.

Most worrying to some investors was the fact that Enron hadn't been a total sham. It owned plenty of real assets and businesses, many of which survived the fall. It had been riven by bad behavior and bad actors, like the CFO Andy Fastow, who went to prison for his scheme of creating and secretly running complex business partnerships to do business with the company and to enrich himself. Such fraud was rightfully uncovered and prosecuted, but there was much about how Enron ran its business units and balanced its books that was perfectly legal. That had some people worried about the risks of fraud that could be hiding at other companies.

GE wasn't Enron, but it used some tricks that were less than transparent. And like Enron, GE was a mystery to many investors and analysts. The company could explain some of its operations, but much of the company was a black box, albeit one producing strong and consistent results.

One of those tricks was the Edison Conduit, a massive special-purpose entity that was technically independent of GE and not on its balance sheet. The problem was that the Edison Conduit was controlled by GE and guaranteed by GE Capital, meaning that GE carried all the risk, while investors didn't even know it existed.

A conduit is a structure commonly used to sell bonds called commercial paper — a form of short-term debt (as brief as a few days and no longer than nine months) that is typically purchased by corporations with excess cash looking for a short-term return. Companies invest in commercial paper to fund their operations. In a large company, money is coming and going but not always at the right time or in the right amount, so access to quick cash is essential to paying suppliers, meeting payroll, and so on. Commercial paper is considered easy to access, trustworthy, and essential. And buyers of commercial paper through the

Edison Conduit were further reassured by the backing of GE and its tri-ple-A rating that these deals were trustworthy. Meanwhile, the mix of assets in the Edison Conduit produced earnings that GE used to repay the commercial paper to investors as it matured.

The Edison Conduit, however, had another, more important pur-pose: it was used to buy assets from GE Capital at prices higher than the recorded book value, creating gains that could be used to improve earn-ings. Of course, GE wasn't transferring the risk of owning the assets, so it was really selling them to itself to produce earnings. But at the time, this wasn't an accounting violation.

In Stamford, Connecticut, the team behind the Edison Conduit took pride in the scheme: a massive earnings engine that was unknown to investors, who saw only steady profits without the risks going into pro-ducing them. Because GE backed the mechanism, GE itself would have to pay down the commercial paper it was selling to investors through the conduit — possibly billions of dollars in cash — if something went wrong. But inside GE Capital, this prospect was unthinkable. They were too smart to make such a miscalculation, and GE's superb credit ratings ensured that it would always have access to cash.

GE had many special-purpose entities that weren't on its balance sheet and were conducting transactions with the company or each other. As required, GE began dismantling these operations as post-Enron ac-counting reforms took hold.

In its 2000 annual report, GE asserted that Capital wasn't a vehicle for off–balance sheet financing. The next year, in the wake of Enron's collapse, the annual report disclosed more than $55 billion of assets in such special-purpose entities. It warned that credit rating downgrades could force it to pay a similar amount to cover issued commercial pa-per. It also revealed that GE's asset sales to the entities brought in $1.3 billion that year. When GE finally had to bring these off–balance sheet entities onto its books in 2003, they added $36.3 billion in assets along with a disclosure that the entities would conduct no new business. GE wasn't so reckless as to take massive gains that would require more and more deals to feed profits. As assets were sold for gains, into the Ed-ison Conduit or elsewhere, the company could offset those gains with nebulous restructuring charges to offset the gain, effectively eliminating the gain. To most entities, a surge in profits would be welcome, but not

to a public company whose results are compared to the previous year. One strong year and a few weaker years doesn't look great to shareholders — it is much better to spread the profits around more consistently to give the impression of a well-oiled machine. That impression created by GE was a true one, but not in the way that people thought. This is the essence of earnings management.

CFO Dammerman acknowledged this practice of offsetting such windfalls to neutralize a big gain from an asset sale. "We have a pretty consistent record of saying, 'Okay, we're going to take these large gains and offset them with discretionary decisions, with restructurings,'" he told *Fortune*.

GE mastered the skill of matching accounting charges with similar gains. Sometimes it would roll in expected costs from the next year to build in even more cushion to future profits. The company pushed back against the suggestion that it was doing anything untoward. They denied that the company was using a "honeypot" — a stash of cash used to hide poor performance — but the scheme's tricks had the same result.

GE's pension funds provided another important tool for increasing earnings. Companies determine how much to set aside for future pension obligations by estimating the return on reserved funds and working backward to a present value. For example, if $1 million is due in ten years, a company needs to put aside far less cash now to ensure that the value increases to the required level. By increasing its expected return on the money in the plan, the company could seem to need less money now to meet future obligations and it could record that surplus as profits. GE boosted its estimated pension returns even in weak years when such outcomes seemed unlikely.

The company also put its own stock in the pension plan, where its high-flying shares increased the investment gains, producing more profit to sustain the stock price. It was a positive feedback loop.

Although the maneuvers involved in managing earnings sound devious, and there is no doubt that at GE they weren't transparent to investors, the practice was largely in line with accounting and governance principles.

Fortune's Carol Loomis wrote that, when she once told Welch that the smoothing practice was terrible, he vehemently disagreed with her.

"What investor would want to buy a conglomerate like GE unless its earnings were predictable?" he asked.

But the winds were shifting. The fall of Enron and other accounting scandals prompted reforms in the rules about the information that public companies had to share with investors. Shareholders were the owners of these corporations, and they wanted to know what companies were doing. More transparency was inevitable for companies like GE. The chairman of the US Securities and Exchange Commission (SEC), Arthur Levitt, warned that at companies that seemed to be making poor decisions in order to meet profit goals, "wishful thinking may be winning the day over faithful representation."

In 2002, the Sarbanes-Oxley Act was signed into law, bringing sweeping changes to corporate governance and internal company controls. It enacted more accounting scrutiny, increased disclosures, and higher penalties for fraud, and it also required that executives vouch for the results they reported. GE's ability to manipulate its reported profits was greatly diminished.

As one GE board member said, "The worst thing to happen to Jeff wasn't 9/11. It was Sarbanes-Oxley."

8

SENIORITIS

Jeff Immelt's assignment was clear: keep the earnings machine of General Electric humming steadily along, as it had under Welch, thereby showing that the company's performance was bigger than one man. But Immelt and his circle of advisers also knew that he'd have to keep GE cranking along without some of the advantages Welch had enjoyed, especially the financial tricks that had made Capital such an effective tool for smoothing out rough quarters and delivering easy profits.

The years of unquestioned adoration were now gone, and new whispers about GE's financial engineering were beginning to be heard. Being the boss of GE was getting harder just as the new boss was still getting his feet under him.

So Ken Langone, a member of the General Electric board, was surprised when Immelt regaled the boardroom one day with what he said was a brand-new priority. Immelt wanted to form a group of his fellow high-profile CEOs to discuss the pitfalls of the CEO job and the unique problems of the most powerful CEOs. Perhaps they could all benefit from the shared scrutiny of their peers.

Langone, the billionaire entrepreneur who cofounded Home Depot, couldn't believe what he was hearing. A new leader needed to be focused on the company — especially one that faced huge economic headwinds — not hobnobbing with other CEOs.

The gruff elder statesman took the new CEO aside. "I see what you are doing," he told Immelt. "With all due respect, I think the time you

spend on this would be better spent with some of the high-potential leaders in the company or with customers."

Immelt listened, letting Langone finish, then fixed him with a stern look.

"I know exactly what I'm doing," Jeff Immelt said.

Langone winced. *Oh shit*, he thought to himself. *Two months in and he has senioritis.*

The world in which Jeff Immelt had thought he would be leading GE had been turned upside down. The recession and the uncertainty that followed the terrorist attacks had dampened the global growth on which GE's industrial businesses depended. And changes to accounting rules in the wake of the Enron scandal, by requiring that the company now account for the vast financial holdings on its balance sheet at GE Capital, had eliminated an easy and reliable source of paper profits to smooth over rough periods.

Immelt knew that he needed to find pockets of economic growth to solve the first problem. He would have to find new markets for GE's heavy machinery and fund the development of new engines, turbines, and healthcare equipment so that sales would get the company moving again in the United States. He also needed to convince investors who had once swooned for the company that GE was no Enron.

Part of Immelt's problem was a small but growing movement to reconsider Welch's success. How much of his nine years of double-digit earnings growth had simply reflected the careful management — if not outright manipulation — of the numbers? And if GE had been relying on accounting tactics that the downfall of other companies now called into question, was GE's aura of success for real?

This was a major and drastic shift. The company had enjoyed broad trust from investors for years, winning plaudits for its investor relations and special notice from one trade publication in the 1990s for being "shareholder-oriented." Through that period, GE didn't even hold conference calls to discuss quarterly results, as competitors did. Its earnings releases were typically short and light on detail. If anything, GE tended to leave investors with questions about how the company had hit its numbers. But the performance of the stock left little room for complaints.

By all appearances, most GE shareholders were happy enough with

their returns to not care too deeply about how the company had achieved them. The stock returns soared, shares split as value rose, and the dividends poured in — a small reward every quarter for being one of the millions of owners of the most famous American industrial company. It was only with the implosion of Enron — its suddenness and the shock of the fakery revealed — that many investors began to think more skeptically about all those smooth quarters and to wonder whether the roaring success of the Welch years could really be sustained.

Welch still loomed as an avatar of success, even as Immelt and his lieutenants felt the growing pressure to prove that the company Welch had bequeathed to them was as solid as advertised. The way out, for Immelt, was to find GE growth of his own.

The problem was that he had little experience in doing this. Running the massive conglomerate was very different from being in charge of a business unit. "Every job looks easy," Immelt began to tell his colleagues, "when you're not the one doing it."

Part of the difficulty derived from GE's structure. As division managers, Immelt and his peers had set and tried to beat aggressive sales and profit targets for their units. But the other side of the ledger, making the tough financial decisions facing businesses every day, wasn't part of their job leading business units. Unit leaders didn't have to worry about their strategy for investing capital to ensure growth, or come up with the money to fund development of new products, or build new factories to handle demand. Some of the basic decisions made by children setting up a street corner lemonade stand could be avoided by GE's elite managers. They simply had to request capital from headquarters.

One former GE executive who left to become chief financial officer at another company quickly learned about the holes in his GE education. This executive had never had to think about cash. GE's corporate treasury vacuumed up the quarterly cash made by the divisions and then distributed it as well. His rude awakening came in his first weeks in the new job. Sounding frantic, his treasurer called to let him know that the company might not have enough cash to make payroll. Not making payroll was a totally foreign concept to this executive. He'd had no idea that his job included keeping an eye on the daily management of money moving in and out and around the company.

At GE, managing a balance sheet had been essentially risk-free be-

cause the company—or more accurately, Welch—took care of that. Now it was Immelt's problem to solve for those below him.

This task was nothing like replacing refrigerator compressors. In his first months in the corner office, Immelt told friends, he felt like he could barely breathe.

Immelt knew that the GE stock price was the gauge that would measure his success and set the tone of his legacy, as had been the case for Welch. But Immelt's self-confidence hadn't deserted him: he was sure that the current dips in the price were temporary. He just needed to sell investors hard on his vision. By early 2002, Immelt was preaching a two-pronged mantra: big changes were coming that would turn General Electric in the direction of new sources of profit and growth, and the company would be throwing open its books as never before to reassure investors that its results were real.

"If the annual report or quarterly report has to be the size of the New York City phone book," he told the *Wall Street Journal,* "that's life."

One day in March 2002, Jeff Immelt looked around the giant auditorium at Crotonville. It was filled with 120 stock market analysts and investors, gathered at the company's invitation to hear Immelt and his team introduce the new measures to ensure greater transparency. But one of the most important guests was missing.

Pacific Investment Management Company (PIMCO), the biggest bond investor in the world, managed assets for sovereign wealth funds and central banks. It was the type of institution whose opinions could sway whole markets. That day, neither its famous cofounder, Bill Gross, nor anyone else from PIMCO had shown up at Crotonville. Jeff Immelt would soon learn why.

Gross's research notes were legendary: his rambling musings on life and human nature were dotted with occasionally sharp judgments on the attributes or failings of companies, politicians, and executives. The note that Gross posted on the PIMCO website on the morning of March 22 smacked GE and its leaders, past and present, right between the eyes. He'd sold $1 billion worth of GE's bonds, he told readers, and his reason for doing so? "The corporation's honesty remains in doubt."

Gross attacked the company on the very issue that Immelt had hoped to boast about: its transparency, specifically with bond investors like

PIMCO. Gross scolded the company for its recent sale of $11 billion in debt, a sale that was followed three days later by an announcement that it was preparing to sell $50 billion more. That announcement had instantly lost huge sums of money for the investors who had snapped up GE's first offering. And the ineptness of the bond sale, Gross argued, indicated deeper problems at the company.

"GE has been shrouded in mystery for a number of years," Gross told CNBC. "Institutional investors have wondered why a company can continue to produce 15 percent earnings growth year after year, quarter after quarter."

The answer, in Gross's telling, was that the success might not be real.

GE's stable earnings were a product of acquisitions, he argued, mostly funded with company stock and commercial paper. Gross, who managed $250 billion in assets, said that he wouldn't buy GE's long-term debt because the yield wasn't high enough for the underlying risk that all of Welch's successes might turn out to have been illusory.

And then Gross hurled a grenade at the commercial paper, the short-term bonds.

The man known as "the Bond King" wasn't going to buy short-term bonds from GE, he said, because they weren't sufficiently backed by lines of credit. GE was selling short-term debt based on the trust of the market. If the market froze or lost trust in GE, it would be out of business because it had no other way to pay its bills.

Gross's comments were a shot at the heart of GE Capital, which generated its outsized returns by borrowing huge sums via short-term debt and commercial paper, then lending it out at higher rates for longer terms. At the time of Gross's attack, Moody's reported that GE Capital's short-term debt, covering commercial paper and debt due within twelve months, totaled a stunning $127 billion and that only 24 percent of this amount, or $31 billion, was backed by bank lines. That $31 billion was the only credit that GE, "the largest corporate issuer of commercial paper in the world," could draw down to pay off the short-term bonds if it couldn't roll them over. In Gross's view, there was a massive and dangerous mismatch between what GE owed and what it could pay back should a crisis hit. GE seemed to be exempt, he pointed out, from the financial parameters demanded by investors in other companies.

Gross's words proved once again that they could move markets. GE

shares slumped as Wall Street worried about what was in GE's black box and whether the company could protect itself from an unexpected shock to the economy or financial system.

But not everyone agreed with Gross. GE was riding out the post-9/11 recession and poised to enter new markets. Immelt and the massive company investor relations machine now assured investors that these new markets — security, water, energy, finance — would surge as the world adjusted to the new era. Gross, in this view, was being paranoid about risk and forgoing investment in a GE Capital machine that reliably churned out money.

Immelt launched a campaign of his own. The CEO appeared on CNBC, where he rejected any questions about GE's accounting and boasted that the multivariate conglomerate was recession-proof. There were always thriving arms of the company that could carry any one that happened to be struggling. GE posted a lengthy online rebuttal to Gross, rejecting his analysis, and took out a full-page advertisement in the *New York Times*. In one-on-one conversations and group meetings with investors, the company reassured the concerned. Gross was wrong. GE was strong.

Through it all, Immelt kept grinning. At one point he joked that he might be wearing out his welcome on television, with his innumerable appearances to talk up and defend the company. Privately, a top GE executive admitted that the company had badly misjudged the level of transparency that its investors would demand. They had work to do if they were to extinguish the doubts that Gross had sparked.

9

LAST CALL

THE PEOPLE OF GE Capital kept a wary eye on Jeff Immelt.
The division's financial profits had fueled two decades of furious growth at GE. Welch had known how central the Capital machine was to his company's performance. Though his own financial knowledge was deep, Welch had also understood his limits when it came to the intricacies of the business. He had entrusted his all-important earnings machine to a trusted CEO, Gary Wendt, and his CFO Dennis Dammerman. When he took over, Immelt would stay fully briefed on what they were doing.

But to the people within GE Capital, Immelt was still an unknown quantity. He had never worked in their division, and from the sound of things, he seemed less interested in finance than in marketing big ideas out of the industrial businesses and expanding GE to other corners of the world.

Welch had chosen a salesman, not a banker, on purpose. Immelt's sales chops, the company believed, would position him perfectly as GE moved deeper into a new century that would require aggressively ferreting out opportunity in new markets as global trade barriers fell, opening new frontiers. But some of those watching from the perspective of GE Capital weren't sure Immelt truly knew how the money got made in the division that had sustained Jack Welch's operation — or if the new boss even really cared.

The focus on marketing throughout the organization led to less risk

management, according to people who worked at GE through the transition. One trend that seemed to take hold inside GE Capital was a greater tolerance for increased risk if it was offset by higher sales volume. The idea was that the impact of a few bad deals would be diluted by the greater number of deals being made. Huge profits would outshine mistakes. As a rule, however, higher investment returns only arrive on the back of increased risk. To old-timers, the idea seemed naive at best and downright dangerous at worst. This type of flawed logic had destroyed many companies, especially in the financial services industry.

The early problems at Capital, including the scrutiny from Bill Gross, and the aftermath of the Enron collapse, along with the slowed economy, led Immelt to take the ax to Capital. He cut ten thousand workers within his first year and completely reorganized the division.

Denis Nayden was pushed out as leader of the division, which was then split into four separate operations: commercial finance, consumer finance, equipment management, and insurance. New leaders in those groups would report directly to Immelt and Dammerman.

As part of his campaign to persuade Wall Street that GE wasn't in trouble or hiding anything, Immelt was now sending it the message that Capital would have more oversight. His assertion that he had long wanted "more direct contact" with the team sent a clear message: I'm the boss now.

The Capital veterans bristled. They were being punished for demonstrating ongoing success while the industrial side of the company continued to depend on their strength. To them, GE had changed the minute Immelt sat in Welch's seat. And not in a good way.

He either didn't understand, or didn't care, that GE Capital was an elite finance organization that was outgunning Wall Street at its own game. It could be stood up next to JP Morgan & Company and Goldman Sachs. This purge only solidified the perception that Immelt wasn't in their corner.

Although he brought a new approach to Capital, Immelt still depended on it. In fact, in his first years as CEO, the operations grew bigger. Like Welch, he found it hard to resist the siren song of the money that could more easily be made from financial services. More than ever as GE's commercial results and stock price sagged, GE Capital was providing the grease that kept the earnings machine running.

To honor GE's handshake agreements with the credit rating agencies, Welch and his team had pledged to limit the company's reliance on Capital, capping its contribution to 40 percent of GE's earnings. Welch himself had no strong desire to set a limit, but the approval of these agencies was critical to the whole enterprise. GE could churn out huge returns only because rating agencies sustained its triple-A debt rating.

Under Immelt, the cap flew out the window. That decision was made partly out of necessity in the early difficult years. More Capital profits were needed to shore up GE's overall earnings. But to skeptics, the rising share of earnings from Capital signified that Immelt wasn't sure what he was doing with a massive financial services business. If too much profit came out of the finance side, GE could face pressure from credit rating agencies to address the risk in its structure. The secret sauce of Capital — receiving a stellar credit rating that let it borrow cheaply because it was viewed as an industrial company, not as a bank — could have been at serious risk.

Immelt's fans thought the criticism was unfair. The CEO was leaning on Capital because of a major drought in the market for power equipment, GE's biggest industrial product line, after years of booming growth. It was slowness on the industrial side, they argued, that made the growth of Capital seem outsized.

At the same time that there was less oversight of GE finance operations, Immelt quietly reduced the board's involvement with financial complexities. In 2002, he disbanded the finance committee of the board, which had been tasked with oversight of "retirement plans, foreign exchange exposure, airline financing and other matters involving major uses of GE funds." He gave no reason.

Indeed, there was no notice of the disbanding of the committee, which had included directors with significant experience managing risk, including former JP Morgan & Company CEO Sandy Warner and the billionaire race car driver and entrepreneur Roger Penske. Listed in the 2002 annual report, the finance committee was simply gone the next year.

Despite all the new transparency, GE still needed to use complex maneuvers to keep earnings buoyant. The playing field for business in the years after fraudulent practices caused the meltdowns of Enron, Tyco, and Worldcom was a perilous environment. New accounting rules took

hold, and more frequent public disclosures were required by the SEC. Immelt and GE chief financial officer Keith Sherin, unlike their predecessors, now had to sign their own names to GE's financials, attesting to their accuracy.

More worrisome to the company than the new pressure from regulators, however, was a growing unease among a group they took even more seriously: investors. Patience for the financial engineering that had papered over short-term hiccups in the industrial business was now wearing thin.

In the fall of 2002, it was dawning on executives at headquarters in Fairfield, Connecticut, that sales and profits were going to come in slower than projected in most of GE's core businesses. In a straightforward accounting of the quarter, the result would have been simple: a slew of expenses and charges were due to hit GE's bottom line. The company would have to report a smaller profit for the three-month period than investors had expected. Almost certainly, some of GE's investors would sell their shares in disappointment and the all-important stock price would go down.

But GE didn't accept such a result without a fight. And conveniently, it had a piggy bank to break: there were gains scattered around the company, including an unexpected tax refund and cash back from customers that wanted to cancel orders. The biggest boost was a $300 million benefit from the sale of GE's e-commerce unit. That deal would give GE the gain it needed to hit its earnings targets and avoid annoying Wall Street, but the bankers were struggling to get the deal closed by the end of the quarter. Into this breach stepped GE Capital. When the buyers couldn't raise the needed funds, Capital bought $235 million worth of bonds to get the deal done.

GE got its gain and hit its targets. It was the sort of gambit that under Welch had earned GE a reputation for creativity and financial guile.

But these moves didn't play that way anymore. What was this, skeptics asked, but the company financing a private investor to take one of its own units away? To the investors who graded GE now, this just looked risky and raised questions about how long GE could make such moves before running out of runway. Immelt simply couldn't haul out the same tools that Welch had used with impunity, at least not without suffering the criticism of analysts and pundits.

Even large investors were raising their eyebrows. CalPERS, the largest pension fund in the country, moved to file a shareholder challenge to GE's executive compensation. The fund wanted leaders to be paid in accordance with their performance.

The party, it seemed, was ending.

10

BUYING AND SELLING

A S GROWTH FAILED to materialize, the price of GE's shares stayed ir-
ritatingly stuck. The incredible growth under Welch just wasn't re-
suming its climb. Whenever Immelt went on TV, he was always asked
about the stock price. He knew it was the most important and most un-
forgiving yardstick used to measure the success of a corporate leader.

In the fall of 2003, GE shares were down 23 percent from the day Im-
melt became CEO. Like Welch before him, the new boss saw the route
to a new era of earnings growth through acquisitions: he would buy suc-
cess by making deals. Wasting no time, GE made three major moves
over three consecutive days in October that showed the company's enor-
mous ambition.

It finalized its $14 billion purchase of the film and television assets of
Vivendi, which would be rolled into NBC to create NBC Universal.

The next day GE closed the acquisition of Instrumentarium, a Finnish
medical devices maker, for $2.4 billion as it sought to move its Health-
care business deeper into technology and patient monitoring.

Finally, the third day brought a deal to buy Amersham, a British life
sciences company, for $9.5 billion, then the second-biggest acquisition
that GE had ever made.

It was a lot to swallow for a company that was showing problems with
its performance and whose young CEO had barely been in the job two
years. A few months earlier, GE had agreed to pay the Dutch insurance

giant Aegon $5.4 billion for most of the commercial lending business of Transamerica Finance Corporation.

To some, the turbo-charged deal-making looked like a series of desperate moves. In mid-October, the *Economist* remarked that "it was not wholly clear whether Jeffrey Immelt, the firm's boss, was climbing out of a hole, or digging himself deeper in."

Strategically, the deals seemed to make sense for the relevant divisions. The Amersham deal would bring the Healthcare division closer to the cutting edge of medicine. It was a long-term bet on the future of medicine, Immelt argued. Still, few could stomach the price he had paid.

GE had shelled out a price that was a 45 percent premium on Amersham's stock price, measured from the day before the news of the deal leaked out. While Immelt and Sherin defended the deal, analysts openly questioned the rich premium. Even worse in the view of Wall Street was that GE had issued shares of its own stock in order to make the Amersham deal in the first place, further diluting a stock that wasn't doing well to begin with and increasing the cost of the dividend paid on each share.

The deal made an impression on a lot of GE watchers. Some board members would later say that it had helped to form a poor impression of Immelt's deal-making skills. The knock on Immelt was that he chased trends, arrived too late, and paid handsomely. One rival CEO joked that he was "fad surfing."

That isn't to say that Welch's record had been unblemished. For instance, some had questioned the real value of NBC to a company like GE. But Welch loved the CNBC network, which was closely watched by CEOs, trading desks, and investment offices all over the country and played no small part in promoting GE to shareholders. Welch personally involved himself in suggesting stories and coverage, calling CNBC a "pet project" and the on-air anchors friends.

Immelt approached deals with a self-righteous resolve. Some of his earliest deals were his best, like the move to snap up the wind turbine manufacturing division of Enron at a bankruptcy auction for just $358 million. It would become the backbone of a major business for GE and give it an important foothold in the renewable energy industry.

But he also made plenty of bad bets. Following 9/11, Immelt began a

push to build up a security division that would sell security cameras and fire detection and related systems for homes and businesses. It never took off and GE later sold it.

Then, in a string of deals, Immelt moved GE into the water processing business. But in pursuit of his dream of profits from building out water filtration infrastructure around the world, the company put its sales force in an increasingly impossible position by having them sell different and competing technologies to potential customers.

Some executives, watching Immelt's frenetic assembly of acquisitions into new industrial business lines, wondered if the CEO misunderstood the meaning and the limits of GE's management magic. The company prided itself on training good managers who could be asked to range across diverse businesses in the conglomerate in order to wring out greater results and improve performance. But the managers didn't walk on water. They couldn't be expected to solve larger strategic problems — especially the question of whether a potential acquisition should be part of GE in the first place. GE's management system couldn't take the place of a strategic rationale.

Immelt had little patience for these concerns. When one company executive who expressed concern about the magpie nature of GE's acquisitions in this period was pushed out of the company over the issue, company veterans were demoralized. Welch had also aggressively bought and sold, moving GE in and out of business sectors, restlessly searching for higher-value operations. In contrast, Immelt's detractors saw his rapid changes as more about inoculating GE from the old and tired and chasing what was shiny and new. Immelt's view was that he was culling pieces of the portfolio that might come to harm the larger entity in some unforeseen way.

There was also a downside to the athletic zeal that made Immelt such a relentless, compelling, and successful salesman. On the buyer's side of the table, once a deal made sense to Immelt, price became less important to him than the story of the strategy and where the deal would fit, his colleagues said. His vision of how a target could be integrated into the company — all the doors to new markets it would open for GE — made price considerations irrelevant, as well as the question of whether it made better sense to walk away. At least that was his rationale. Immelt rarely folded his hand, even when some of his lieutenants thought he

should. To him, leadership was perseverance in the face of doubt. And opposition to that approach wasn't just disagreement but something worse. To Immelt, naysaying in these situations was a form of betrayal.

Immelt wasn't only on a spending spree. He was also selling.

With Welch's exit came the souring of GE's massive insurance bets. Like other companies that had tried to make money in the business, GE had to boost the reserves of capital it held to cover future responsibilities as it became clear that previous expectations had been too conservative. Immelt had to plow $9.4 billion into the insurance units in just his first five years at the helm, to increase reserves.

The issue was reinsurance — the insurance that GE had sold to back-stop other insurance companies. GE had sold this coverage to help companies shrink the risk of their business, with costs pegged to the level of risk that the threshold to make payment would be reached.

A reinsurance policy can protect a company from excessive losses. For the seller, the art lies in pricing the policy correctly to turn a profit. In the meantime, the seller uses the cash from premiums to invest and get even higher returns. GE saw an opportunity.

Welch had snapped up Employers Reinsurance Corporation (ERC) in 1984 for more than $1 billion. Other deals followed as GE dived into a niche that seemed like an easy profit center. Welch later reversed himself, admitting that GE had mismanaged a business it didn't know well enough. It was too aggressive and mispriced what it was selling, a potentially fatal move in the insurance industry. As a result, GE had to put up hundreds of millions in reserves in the late 1990s.

But Welch had loved the business, once declaring that GE's experience in insurance "gave us the confidence we could do more." These GE acquisitions, he wrote, "seemed like easy additions to a good thing." Instead, they exploded under Immelt, who had to simultaneously shore up GE's insurance businesses and then exit the industry. Immelt ruefully joked that GE had no business in an industry that had until recently sold insurance for pets.

Immelt would prove his worth as Welch's successor, he knew, by getting out of businesses as much as by getting into new ones. And he viewed insurance as a business to get out of. He would time it right, just as the market was peaking, and not wait so long that a major finan-

cial crisis could squeeze the unit's earnings. The new boss wasn't wrong. Deeper credit emergencies were coming, even though GE executives, including CFO Sherin, repeatedly declared over the years that the reinsurance risk was gone. Immelt himself, defending the company in a letter to *Barron's*, noted that the company's auditors had signed off on their books. "Each quarter and every year, GE uses a rigorous process to review reinsurance claims and estimate reserves," he wrote.

The company spun off most of its insurance holdings into Genworth Financial in 2004 and sold much of the rest to the Swiss Reinsurance Company two years later. But there was some fine print in the deals.

Genworth had a block of reinsurance that covered a variety of long-term care insurance policies, which cover expenses like nursing homes and assisted living. This coverage was gobbled up by customers, but it was priced all wrong and eventually wreaked havoc across the industry. When preparing for the Genworth spin-off, GE's bankers advised not making the long-term care business part of the offering. So to make the deal more attractive, GE agreed to cover any losses from that group of policies.

In short, the buck stopped at GE Capital. And what was the risk? What could possibly threaten GE's cash machine?

IMAGINATION AT WORK

THE MODERN GENERAL Electric was the house that Jack built, but Jeff Immelt had shown that he planned to do some remodeling. He moved to put his mark on the company, even if only superficially at first.

He painted the corporate jets, a fleet that included a 737 that shuttled GE executives all around the world. Immelt, as GE's CEO, was required by the board to use only the company's planes and barred from flying commercial. The reasoning was to ensure his safety — an excuse also used at other big companies — but protests about such "rules" were rarely heard.

The painting of the jets by the new leader was perplexing to those at the top of the company. Although the painting project wasn't expensive enough to move the needle at the massive company, it did cost a lot of money. Furthermore, the jets didn't need to be painted and the reasoning for the move was unclear. It was another lesson about Jeff Immelt. If he wanted something, he got it and it was best not to question too much.

General Electric's portfolio continued to churn as the company bought and sold pieces of its business lines. Immelt, aware of the concerns of Wall Street and eager to put a more personal stamp on the company, had come up with a new mantra: organic growth.

That was just Wall Street jargon for growing the businesses you have, rather than growing by buying other companies — the inorganic approach that some investors like Bill Gross accused Welch and Immelt of pursuing. Immelt was eager to prove that he didn't need acquisitions to

produce strong profits — that he could supercharge GE by stirring up its existing units to produce more revenue, which would fall as profit right to the bottom line. It sounded good to analysts, though Immelt's plan would work only if he and his team could find ways to bring in new revenue without much additional spending.

The headwinds of Immelt's first years hadn't vanished — the economy remained sluggish, investor scrutiny was still harsh, and stock returns continued to be lackluster — but the promise of steady organic sales growth was a clean path he could promise to take in leading GE into the new century. Taking that path would also be a good reason to focus on one of Immelt's favorite parts of the business: marketing, a facet of the company that he felt could be its differentiating power.

Immelt believed deeply in the power and value of GE's brand. Indeed, the force of its broad recognition and global approach alone, Immelt believed, would help him grow the company's sales. But he also wanted to market the company more aggressively to the media and the public, to show society how impressive the company and its products really were. And he had just the right executive in mind.

Beth Comstock had spent her career in media relations, rising through NBC to the corporate parent, when Immelt tapped her to be GE's chief marketing officer in 2003. The company hadn't had such a position for two decades, but Immelt believed that Comstock, despite her lack of experience in marketing, was a "change-maker" who could reshape the image GE presented to the public and help him drive growth in the company's sales. Immelt was "raising our game" in marketing, he said. Comstock dove into the role, reading marketing textbooks to learn the basics and networking with other CMOs.

In choosing Comstock, Immelt, like Welch, was going with his gut. He wanted more than just an aggressive new marketing executive. Immelt's stated mission to Comstock was to "resurrect marketing at this company." He wanted to put his own stamp on the GE brand itself. The company said that the brand awareness it derived from the stalwart Monogram logo was worth billions. But in spite of its immense value, Immelt felt that the GE brand needed refreshing: a modernization that would better reflect the twenty-first-century image he was trying to put forward.

The new boss thus set his sights on one of the most iconic elements

of the brand: the advertising tagline that GE had used since 1979, familiar to millions of Americans who knew the jingle from GE TV commercials that hawked everything from lightbulbs to clock radios: "We Bring Good Things to Life."

Immelt wanted something newer, breezier, and evocative of the technological innovation he planned to emphasize as GE's core offering. It was lost on no one that his desire for a new tagline would also sever the company more completely from the Welch era, and all the lingering comparisons that arose from it.

Immelt turned to BBDO, GE's go-to advertising agency for decades. The request was no small undertaking. The old tagline and jingle were still inextricably linked to GE in the public mind. The advertising tune itself had been composed by a veteran jingle writer named David Lucas, whose compositions were heard in ads for huge corporations every day. (Lucas's more eye-opening claim to fame: moonlighting as a rock producer. He claims to have been the one playing the cowbell on Blue Öyster Cult's "Don't Fear the Reaper.") Even if Immelt felt that Welch had neglected his marketing responsibilities, the company had still plowed more than $200 million into ad campaigns in Welch's final year, an investment in the trademark that Immelt now wanted to wipe away.

BBDO came back to Immelt with a surprising suggestion: leave the slogan alone. GE would be reckless to make a change, the advertising staff said. GE's brand was the strongest in the market other than Coca-Cola's.

But Immelt wasn't having it. It wasn't an evolution that he wanted, but a revolution. If BBDO wouldn't draw up a better tagline, he retorted, "then I'll change ad agencies."

BBDO followed orders and went back to the drawing board, returning with a phrase that perfectly captured what Immelt wanted for his company's new look — the sense of an organization that could make happen whatever its customers dreamed up: "GE: Imagination at Work."

Immelt loved it. The new tagline summarized the innovation that would come out of GE. It conveyed new ideas and unlimited potential in the brave new world driven by high technology. GE was no longer about lightbulbs, turbines, or engines — it was about dreaming up the next era of world-changing inventions.

In fact, Immelt wasn't making quite the break from GE's past that he

liked to think he was. Welch had once told shareholders that the company's talk of setting "stretch goals" for its performance "essentially means using dreams to set business targets — with no real idea of how to get there." Like the guy he had chosen to succeed him, Welch had believed that an aspirational goal could always be met by a sufficiently motivated and determined worker — or if not by him, then by the colleague who wanted to take his place. Still, the new tagline represented Immelt's effort to declare his transformation — however vaguely defined — of the massive company he now led.

"Imagination at work is not a tagline," Immelt declared. "It's a reason for being."

The overhaul kept going. GE tweaked its beloved meatball logo and offered the Monogram on a solid field of baby blue. Marketing consultants from Wolff Olins squeezed the company's 3,500 business offerings into eleven market groups so that customers could more easily understand what this new iteration of GE made and sold.

GE even commissioned its own typeface, to be used in everything from advertising copy to press releases. The company said that the font was "conceived from" the rounded letters of the GE Monogram. GE named its new typeface "Inspira." (Never prone to underselling, the company described the font this way: "When introduced, the rounded and flowing forms of Inspira caused some to question its open nature. Others felt a breath of freshness and inspiration.")

New slogan in place, an ad campaign followed. BBDO's campaign mashed up images of GE's past and present, making an argument for the company's technology as a force shaping the world. Hard-rowing Vikings surpassed by a sailboat highlighted the ability of GE's wind turbines to harness the wind. At Kitty Hawk, the Wright Brothers' flying machine was now powered by a GE jet engine strapped on top of it. Edison, naturally, popped up in an ad depicting electronic health records that whisked information to doctors in seconds.

Immelt felt that the rebranding effort was a home run, though some in the company grumbled. Was the new boss marking his territory? Was all this branding a solution in search of a problem? Were the splashy new ads superficial? And what did the tagline really mean, especially compared to the old slogan? Everybody knew what bringing good things to life meant.

The eye-rolling was especially pronounced at GE Capital. Whimsical advertisements and talk of imagination were all well and good for Super Bowl commercials. But GE Capital dealt with hard numbers. Given the wider scrutiny of accounting and transparency, including at GE, some at Capital thought that the new slogan was insulting, if not risky. GE Capital saw itself as an elite finance operation — better than the hidebound bankers of Wall Street — and the sturdy scaffolding that was holding up the results of the lumbering conglomerate. As they saw it, there was no imagination in their work. Nor was any needed.

Immelt dismissed such concerns. This was the new GE, he felt. There would be those who couldn't keep up with his changes, but those who did were his fellow visionaries. They understood that the company's story of itself was the heart of its strategy. Selling products wasn't enough. The company was doing more than making widgets — it was making the future. And Immelt, his good cheer notwithstanding, was not interested in hearing his judgment questioned.

"My job is to make the company perform," Immelt told a newspaper reporter, "and my job is to make sure that nobody defines this company other than me."

The doubters soon realized that "Imagination at Work" wasn't just advertising. Immelt was intent on putting marketing at the heart of GE strategy to dictate not just how the company sold the things it made but what it made in the first place. Much as Welch had before them, Comstock and Immelt hatched new jargon to express the process they wanted the company to follow. GE business leaders would now convene to come up with "Imagination Breakthroughs" — that is, ideas about products the company should design and sell.

Immelt wanted division heads to generate imaginative new product and service concepts, which in turn would generate the new organic revenue on which his vision depended. It was a tall order: a handful of product ideas that would each pull in $100 million in new sales for each business. More important, Immelt wanted these "breakthrough" sessions to be led by each unit's marketing department — to have the division that usually dictated advertising and branding stepping into the role that had been the province of product engineers. Immelt's inspiration for the directive was an article he read about a smaller industrial

conglomerate called Danaher Corporation that had formed an internal incubator to develop new ideas that could drive revenues and profits. Its CEO was a young whiz named Larry Culp who, at age thirty-seven, was even younger than Immelt had been when he took the reins.

The marketing focus was clear across the company. Immelt established the Commercial Council, an intracompany panel composed of "the best sales and marketing people in the company," and served as the chair himself. "That turned out to be a big deal," Immelt told the *Harvard Business Review.* "The council was designed to share best practices and plan growth programs, but more fundamentally, it began to develop this idea of growth as a process."

The marketing people were now charged with running GE's growth strategy. This new focus brought a lot of new spending and a lot of advertising. Everything the company did now had to have a story behind it, and all decisions and strategies had to fit into a broader story — or else they made no sense to the new team at the top.

Story was strategy, Comstock would argue.

Spinning that story cost a lot of money. As Immelt changed the master narrative of General Electric for his audience, intent on his strategy for growing top-line sales and reaping profits by keeping costs low, executing his strategy was becoming increasingly expensive. The GE corporate umbrella — the layer of the company that oversaw the individual businesses — began to spread, and it would continue to expand, in size, cost, and complexity, throughout the Immelt era. For the CEO, the true benefit of the conglomerate lay in having all its disparate pieces gathered together under a strong organization to oversee it all.

They called it the "Borg" — the unified research, sales, information technology, and back office departments housed within the corporate hub of GE. Borrowed from *Star Trek*, the term referred to an alien race that assimilates others into its hive mind of collective intelligence, while erasing individuality.

Seizing on a Welch-era concept, Immelt called on GE workers to pursue "boundaryless sales": selling not just particular products from their own GE business unit but the entire catalog. Taken even further, you were now selling the entire product portfolio. An executive selling a locomotive in Delhi was in perfect position to nose around for leads about

prospective buyers of X-ray machines or electrical switchgear or asset-backed commercial loans.

All those salespeople leaned back on the Borg, which funded their jobs as well as the research and production costs of the heavy industrial goods they were selling. The idea was to allow more than half a dozen industrial companies to lean on the same source of financing and an overlapping pool of talented researchers and patents in order to achieve the ultimate sales synergy. It was a strategy perfectly suited, Immelt felt, for a world in which the battle would be fought for market share in the burgeoning frontiers of the Middle East, China, Africa, and South America.

The challenge would be ensuring that profit margins on sales exceeded the growing costs of the Borg — all those layers of marketing, research, finance, and other supports for the industrial operations. But Immelt believed that the promise of future sales would offset the cost of swelling the corporate hub.

Immelt felt that he was truly innovating, forging a new unifying vision that justified the conglomerate structure so out of favor elsewhere. Danaher, for example, used a very thin umbrella, only a few hundred people, to oversee its portfolio companies. Berkshire Hathaway, the conglomerate run by the billionaire investor Warren Buffett, was even leaner, with just a handful of people at the top. All other functions were put into the divisions so that they could manage costs appropriately and then be held accountable for keeping them under control.

By contrast, GE's corporate structure not only could prevent managers from developing the skills they needed to run an independent business, but also made it impossible to get an accurate picture of a division's performance or health. Measuring profit without including all the underlying costs distorted the performance of the businesses and their management. It was great for painting a picture of success to investors. Less clear was whether the thickening of the corporate layer was a solid strategy for the long term. Immelt was effectively undoing the aggressive cuts that Jack Welch had inflicted on the GE bureaucracy. Welch's dreaded layers were now returning under another name.

One more element of Jeff Immelt's vision would now dictate GE's strategy. Tying the company's fate to growing top-line revenues required finding markets where sales were growing. For power equipment, loco-

motives, jet engines, and hospital machinery, the critical markets were not likely to be found in Europe or the United States, where GE had historically focused. The company's footprints had been laid around the world for decades, but they were mostly just that: footprints. Immelt's notion of a wave of sales growth that would alter the makeup of GE's earnings for good depended on finding major new sources of sales outside the West. In short, GE would have to go to China.

GE wasn't the first company to boldly move into international markets, but it faced unique challenges in doing so. Tapping international markets wasn't as easy as just opening an office in a foreign city; it required deep connections with local, regional, and national governments. Moreover, GE wasn't selling cell phones, or any other product that could be sold just by parachuting in. Its machines cost hundreds of millions of dollars and would be in place for decades. GE needed local operations that understood their market and could establish roots.

Entering new markets, including those that weren't stable or particularly welcoming to American interests, could also get complex and risky. On some sales trips to unstable regions, Immelt would sleep in the corporate jet because it was considered to be the safest place for him to stay. It was often difficult to win business in countries saturated with corruption, especially for GE, which claimed that it was the rare multinational company that played it straight.

Growth at all costs, grabbing market share, and establishing a presence were all essential to selling in frontier markets. But struggling to profit while expanding was okay with Immelt, because the spoils would come later, he hoped, as GE leveraged its size and new customer relationships in those markets.

Immelt traveled almost constantly, spending about a quarter of his time, he told associates, outside the United States. He had a travel schedule and a retinue befitting a world leader. And there were signs that the world reciprocated. Amid a war of words about trade tensions between China and the United States, Premier Wen Jiabao stopped in New York City in 2003 on his way to meet with President George W. Bush. Wen and his entourage made their way to 30 Rockefeller Center, where they endured the full dog-and-pony show — demonstrations of a locomotive customized to Chinese working conditions, a jet engine for cross-Pacific flights, and a security scanner for better crowd control in advance of the

country's Olympics. Wen's government had already signed on for $900 million worth of GE power turbines, and it had also offered the company a slot building engines for China's first commercial airliner, which GE said would generate $3 billion in profit over time.

"In my view the cooperation between Chinese enterprises and GE is successful," Wen told the Associated Press before heading to Washington to meet the president.

The world had finally heard the new story, it seemed, about Jeff Immelt and General Electric.

12

TIME TO GO BIG

THE STOCK MARKET didn't appreciate what General Electric was really worth. And it was driving Jeff Immelt crazy.

His handlers claimed that he didn't watch the daily movement in the shares, but his actions betrayed him. The stock market was the ultimate scoreboard tracking his performance. But rather than addressing investor concerns about the company's results, future growth, and transparency, he instead gave a sales pitch.

"The stock is currently trading at one of the lowest earnings multiples in a decade," he wrote in his annual letter to investors in early 2006. GE had "a great team generating record results with more valuable businesses that have high visibility on future growth."

"Investors decide the stock price, but we love the way GE is positioned. We know it is time to go big!" he wrote.

Immelt possessed a legendary ability to put people at ease. His cool confidence telegraphed that he knew what he was doing. When challenged, he used his sharp wit and well-worn talking points to explain why he was right. His loud guffaw and easy backslapping lightened moods and closed deals.

But that wasn't working on Wall Street. Not anymore. Smooth-talking leaders from Enron, Worldcom, and Tyco had left a legacy of bitter wariness that would not soon be forgotten. And the financial models of sophisticated investors simply didn't factor in Immelt's optimistic visions

of where GE was headed. Performance was needed to move the stock, not promises.

That didn't stop Immelt from trying to argue the stock price higher. Sometimes, when Immelt tried to base his case on the numbers, he left Wall Street scratching its collective head.

He proclaimed that GE would deliver sustained organic growth — that is, bring in more money to its existing businesses, without acquisitions — of two or three times the growth rate of global gross domestic product. In 2005, global GDP grew by 3.8 percent, according to the World Bank. The notion that GE could double or even triple such growth over the long term wasn't just optimistic: to many observers, it sounded impossible.

"No company has ever achieved the kind of growth GE is seeking, and certainly not on a revenue base of $150 billion," wrote the *Harvard Business Review*, which also noted that the path to that growth wasn't clear.

Nevertheless, Immelt stuck to detailing his strategy only in broad, vague strokes. One element of that strategy was having employees pitch ideas to him for approval. He was serving as the "program manager" so that workers could "live their dreams" on the platform he was providing.

"Size facilitates growth. It gives us the chance to take swings, spread successes, and learn from failure," he wrote in the same shareholder letter. "We have developed a large pipeline of Imagination Breakthroughs that can create growth in the short and long terms."

He projected that GE would generate more than $10 billion of free cash flow per year and explained that this was cash GE didn't need. It would be what was left over after paying dividends. "Our industrial businesses do not require much investment to grow, while our financial services businesses have inherently high return on equity," he explained to shareholders. Pledging another $25 billion share repurchase plan, he assured them that "the next few years should be good ones for your Company."

In 2006, GE's stock rose 5.2 percent while the S&P 500 rose 15.6 percent.

• • •

In Norwalk, where GE Capital was based, they weren't dealing in dreams, counting on imagination breakthroughs, or taking big swings at transforming the company. The way Capital and its people saw it, they just made money.

The search for returns was never ending. Capital was constantly on the hunt for deals to finance, assets to buy, and businesses to run. From taco stands to ocean freighters, they were putting GE's money to work.

But all those years of success at Capital had also laid a trap. As big investors and other financial companies had unpleasantly discovered, the bigger Capital grew, the harder it was to move the needle with a successful deal. The unit could get higher returns only by shouldering more risk. Capital had little choice but to go after acquisitions or new lines of business that it might earlier have rejected.

Capital's massive size also enabled it to jump into new markets in a significant way, making major plays with the hundreds of millions it had to put to use every day to make a profit on the back of GE's sterling credit. Subprime mortgages were just such a market in 2006. A white-hot housing market had set off a flurry of lending, with banks and mortgage lenders cranking out mortgage loans even to borrowers whose credit rating was poor — or not even fully known.

For GE Capital's executives, whose marching orders were to invest the corporation's money and put it to work, leaping into the subprime market wasn't a hard call. And there would be little objection that an industrial company that specialized in selling and servicing giant machines should think twice before entering the mortgage market. Capital was already a vast financial operation, and the almost absurd breadth of GE's businesses was well established. It was an industrial company that also broadcast *Friends*, sold insurance, and had a long line of previously owned properties that ranged from Gibson Guitars to a massive Australian metallurgical coal mining business to the department store Montgomery Ward.

Why would it be scared to add some portfolios of home loans?

GE Capital had taken the plunge in 2004 by buying WMC, formerly Weyerhaeuser Mortgage Company, for about $500 million from the private equity firm Apollo Global Management LLC, the lender's owner for the past seven years. Apollo had purchased the lending unit from Weyerhaeuser, a timber company that was one of the largest private

landowners in the United States. At Weyerhaeuser, management had decided to simplify its portfolio, selling off the lending business to concentrate on its core business. GE was headed in the opposite direction.

For Capital, it was easy money. WMC lent money to home-buyers, then flipped the mortgages, selling them to investment banks that wanted to package the streams of loan payments into bonds. WMC, and thus GE Capital, got its money up front from those banks and went back to work finding people who wanted to borrow.

All the assets on GE Capital's balance sheet had been worth less than $200 billion in 1995. That still made it a bigger financial unit than any modern industrial company had ever operated, but one that the leadership said was just as safe and manageable as the rest of the company's portfolio. It had been supercharged in Welch's years, growing to $425 billion by the time Immelt took over, as GE leaned on the unit to keep generating profits and smoothing returns. By the end of 2006, Capital had mushroomed, having grown by another third, to $565 billion in assets. And it wasn't done growing.

GE Capital wasn't only minting earnings. It was also producing executives. Among them was a short, gruff, hard-charging native of Lewiston, Maine, who had risen through GE's legendary corporate audit staff to become one of the key lieutenants to the men who ran the lending unit. He loved to fish and had a theatrical disdain for dumb questions. His name was Jeff Bornstein.

Bornstein had come to GE after studying finance at Northeastern University in Boston, proudly lacking the Ivy League pedigree of many colleagues. As an employee in the Power division, he had joined the rigorous financial management program. The two-year program amounted to an MBA of sorts for young GE recruits, and it put those who completed it on a fast-track rotation through different roles at divisions across the conglomerate's businesses. It was the first step for people whose ambition was to rise to the highest levels of the company.

From Power, Bornstein moved on to the Corporate Audit Staff (CAS). This finance program, another elite job rotation, was the Praetorian guard of GE's corporate culture and a pathway upward for the most ambitious of GE's young executives. It traced its roots to the early twentieth century and accepted only a couple hundred workers a year. Most

of the top tier of GE's management were graduates. CAS was a notorious burnout machine: participants were expected to sublimate virtually everything else in their lives — their schedules, family obligations, outside interests — to the company and the demands of their assignments.

CAS teams dropped in on GE businesses and audited their finances, operations, and anything else relevant. To get a broad view of the company, the teams rotated into different GE businesses every few months. The CAS role, with its policelike authority over the books of other units, came with a lot of power. Having the head of the program report directly to the chief financial officer turned CAS into a de facto network of information about the inner workings of various business units available only to the CFO and a very few others across GE.

CAS also worked closely with KPMG, GE's external auditor for more than a century. The accounting firm often relied on the findings of the staff to guide their own work in reviewing GE's endlessly complex finances. Indeed, the relationship was cozy, given GE's habitual forays into aggressive, questionable accounting practices. To outsiders, like reporters and market analysts, GE leaders referred to CAS as an additional layer of independent auditing, serving a function much like the one KPMG was supposed to serve. In practice, there was little to separate GE's audit staff and the company's interests. If anything, CAS was as responsible as some executives at the businesses they examined for digging up new opportunities for profit.

If KPMG rarely questioned GE's findings, the corporate audit staff was even less prone to uncomfortable demonstrations of independence. In fact, they were incentivized to juice GE's accounting wherever possible. CAS teams descended on businesses to make sure they were performing, but also to search for ways to adjust practices to increase profits. In other words, the crack internal squad with the word "audit" in its name was also being paid to search for ways to boost GE's quarterly earnings.

One fast track to getting the attention of GE's CFO was to win the Corporate Audit Staff Honor, an award that went by its acronym: CASH. The honor typically went to staffers who had found large beneficial accounting adjustments — ways to tweak the numbers, within the bounds of the law, that boosted the profits that GE would report to Wall Street.

The CAS program consisted of two two-year increments, at the end

of which successful entrants could become bona-fide company executives after just five years. The burnout rate was intentionally high; "up or out," the program rules said. After most of the ambitious candidates were bumped out of the program, only a handful were left who made it all the way through, having been invited back for each level of the program. Still, a decent showing in CAS, even if bumped in the end, could be the start to a solid GE career. Lists of those who hadn't been awarded another year at CAS were passed around to HR executives in the industrial businesses, who used it as a sort of waiver wire for workers who had been almost good enough to be on the inside track to GE's highest echelon of leaders.

Jeff Bornstein made it all the way through, exiting with the internally propitious title of executive audit manager. He also managed to meet his wife Jill in the process. She also went through both programs and landed a job in credit card finance.

From there, Bornstein's stock soared within the company. In 1996, he became CFO of GE Aviation, and then he became a GE officer two years later. At just thirty-three years old, Bornstein was one of the top two hundred executives in the company. His meteoric rise made it clear that he was being groomed for one of the top slots in the corporation.

When Immelt split GE Capital into four parts reporting directly to him, Bornstein was installed as chief financial officer of the commercial finance unit. The business was the largest piece of GE Capital and managed assets of about $180 billion.

Then, in 2005, Bornstein was named chief financial officer of GE Capital. He joined the small group that gathered for lunch most days outside the office of Mike Neal, the CEO, on the top floor of the office in Norwalk. Bornstein and Neal became close and would retire together to the roof to smoke cigarettes and talk about business, away from the eyes and ears of others.

As the decade rolled on, the pair pushed GE Capital to get bigger and bigger through the relentless pursuit of more deals. Capital bought $2 billion in commercial finance assets from Boeing, which was selling in order to focus on financing only its own products. Unlike GE, Boeing was no longer interested in being a bank.

Capital then bought 40 percent of Hyundai Capital Services and Dillard National Bank, which was a private-label credit card business. It

bought Bombardier's inventory financing division, and it spent $1 billion on aircraft assets from CIT Group and $1.8 billion for a 25 percent stake in Turkey's third-largest bank. Capital acquired Antares Capital in late 2005, adding an important puzzle piece in the middle-market financing world that worked with private equity funds.

There was a swagger to GE Capital's deal-making. It acknowledged no limits or borders as it rolled up assets that others didn't want. Size and volume were only benefits. GE had a seemingly bottomless pot of cash to put to work. Its deal-makers — its salesmen, as one executive called them — just had to go find places to put it.

They went, eventually, to the end of the world. In 2006, GE bought the $500 million mortgage portfolio of New Zealand's Superbank, a joint venture that aimed to bring mainstream banking to supermarkets. The unusual operation would last just three years. But GE had what it wanted: more assets on the balance sheet — another heap of business on the GE Capital pile.

13

HIGHER RETURNS

As capital continued to grow, the old-line industrial conglomerate in which Jeff Immelt and his lieutenants had come of age was changing. Such change was a GE tradition, in many ways. Jack Welch had impatiently cashed GE out of industrial businesses as he tired of them, especially those that represented the big bets of a previous generation of leadership. He had moved quickly upon taking over as CEO to get out of its massive mining operations, which included a major footprint in Australia as well as exploration rights across the American West. Welch saw little advantage for the company in resource extraction plays. He had become similarly fed up with some businesses that were getting commoditized, like the small appliances unit. In the 1980s and '90s, it was no fun, and not good business, to try to keep up with the Japanese selling clock radios and toasters.

Immelt was doing a similar round of examinations of the GE portfolio. With an eye toward the swift-moving globalization that was opening up supply chains across oceans and upending manufacturing businesses, he was especially wary of GE product lines that became commoditized, as the small appliances had been. Where he thought investment in overhauling the product offerings could boost earnings, Immelt made new investments, chasing after the market for new products like compact fluorescent lightbulbs and the zone-line air conditioning units the company sold to hotels.

Immelt was skeptical about those industrial units whose salespeo-

ple competed the hardest on unit price. The best jet engines in the world could be sold at a million-dollar loss; the incredible expense of designing, building, selling, and installing the engines became worthwhile when the second and third maintenance overhauls of those engines came due a decade or so later. Then they gushed out pure profit, just as a power turbine did. But that wasn't how the washing machine business had ever worked. And now it was under assault from increasingly attractive and low-cost machines from new manufacturing powerhouses like South Korea, and facing a consumer class with whom GE's old promises of durability and familiarity went nowhere.

In a culture in which people were accustomed to tossing more treasured devices like phones and computers after just a few years, no one expected to get a full generation of use out of a dryer anymore. Neither did they find national loyalty to be a compelling reason to buy an American washing machine if a Samsung machine sold for 20 percent less. Channeling the cold-bloodedness for which Welch had been renowned, Immelt wanted no part of a business that relied on outmoded expectations. The company quietly began looking around for buyers for the massive Appliance Park business in Louisville, the factory complex in which GE had supplied, and helped to create, the post–World War II American middle-class home.

Some other pieces of the portfolio were more quickly disposed of. Immelt sold specialty materials assets in Europe, the silicones business, and the advanced materials business to the Apollo Group, a private equity firm, for $3.8 billion. In the announcement, GE vice chairman Lloyd Trotter said that the deal would help the GE industrial businesses achieve "faster growth and higher returns."

An even more storied piece of the conglomerate came next.

Plastics — the training ground for Welch, Immelt, Bornstein, and countless others — was also sold. Formed in 1930 and central to the prosperity of Pittsfield, Massachusetts, the division had changed the world with the invention of Lexan. Its plastic pellets were turned into products all over the world. It made the plastic cases for Apple's famous iMac G3, the multicolored desktop computer that had heralded the triumphant return of Steve Jobs as CEO.

Immelt, true to form, had poured money into marketing at Plastics in an attempt to make Lexan and other products into household names.

The Plastics marketing team wanted consumers to identify its product lines, much as Intel had succeeded in teaching the broader public about its computer chips by using "Intel Inside" stickers as a sign of quality.

In an early intracompany experiment with the internet, GE set up a site it called "Polymerland," which sounds more like an ill-fated amusement park designed by chemical engineers. The site was intended to facilitate business-to-business sales of the plastics needed to produce goods like appliances and power equipment.

It hadn't made enough of a difference. And the dwindling margins were not going to be propped up by even the best internet-era buzzwords and product stickers. Oil prices in particular strained the cost of the raw materials needed in the Plastics division. Knowing the score, Immelt cut it loose. GE Plastics was sold to Saudi Basic Industries for $11.6 billion.

Immelt said that the sale would aid GE's "faster-growing, higher-returning set of businesses capable of delivering sustained performance." Of course, the cash would also be used to repurchase lots of GE shares.

Amid the mad deal-making and growing dependence on financial services, Immelt constantly pledged to investors that GE knew what it was doing. He assured them that the solid blue-chip company of the twentieth century was as dependable as ever, but was now blazing an innovative new path to lead the way for other companies of the future.

Risk was being well managed. In his 2006 letter to investors, Immelt boasted that GE managed more than $560 billion in financial assets "with losses less than the industry average." He praised the oversight of Capital's monthly board meetings, which included a memo written by the chief risk officer, Jim Colica, giving his views on every deal. Immelt said that he also spent an "hour alone" with Colica "to review each deal through his eyes."

Yet the internal detractors remained. Others described a GE culture so different from Immelt's public pronouncements that they questioned whether he really understood how Capital worked. One thing was certain: GE regularly leaned on the group to make sure that profits stayed steady.

14

APPLIED MATHEMATICS

Few fates were worse than missing your numbers at General Electric.

The conglomerate was performance-driven and had a thick top-down management structure. Executives assigned targets to underlings, rather than lower-rung workers passing information up the ladder, so projections were based on market realities.

Managers were expected to hit those targets at the end of every quarter, and anyone who came up short scrambled to cover their required contribution to feeding Wall Street's expectations. Since the Welch years, GE had rarely missed a quarter's projections, and that wasn't an accident.

GE Capital's enormous base of assets provided the rest of the conglomerate with plenty of chips to cash when needed. If the quarter was looking light, something could simply be sold — a building, a parking lot, an airplane, or anything else in the cornucopia of GE holdings — for a quick profit.

For the divisions, the work was a little more intense. As one former senior executive told it, there often wasn't much downtime after Thanksgiving, as long hours were required to make sure everything was getting booked by the end of the year.

Some GE workers didn't spend New Year's Eve with family or friends. Instead, they were at their desks in half-full but bustling offices. Food

and libations were brought in, but not for a party: workers needed to be there to process last-minute changes and orders.

The end of the year was a time to push. The Healthcare business might need to make an eleventh-hour call to a hospital CFO to make a quick sale of an MRI system or other equipment, priced to move. Sometimes late sales like this created a mad rush to get the ordered equipment out on a truck so it could be recorded as a sale before midnight struck and the accounting year officially ended.

In GE's entertainment division, staff accountants tracked big-budget movies to see if they were living up to projections, especially as the all-important quarter drew to a close. Movies produced by Universal were assets as far as GE's accounting was concerned, merely investments on a balance sheet. Accountants totted up the expected value of a motion picture before it was released in theaters, subtracting the cost of production from the box-office revenues they expected to take in. But if a bad review, or slower than anticipated ticket sales, reduced the value of an asset, that change to the estimated value of the film the company owned had to be reflected on the books.

That had been the worrisome fate a couple of years earlier for *King Kong*, a remake of the Hollywood classic that had a star-studded cast and was directed by Peter Jackson, fresh off his success with *The Lord of the Rings*. The studio had expected a blockbuster, and they had spent like it. With a budget of more than $200 million, *King Kong* was one of the most expensive films ever made.

Critics were kind, but audiences were not, and the movie, while profitable, was pulling in far less than GE had estimated when it assigned a value to the film on its books. To the accountants, the carrying value of the movie needed to be adjusted, a move that would hurt GE's profits. When the finance staff warned their bosses about the charge, they got an unexpected response: don't make the adjustment.

The group was in a pickle. They couldn't simply pretend that the movie was going to suddenly meet its gargantuan expectations; they knew that the number would have to be adjusted downward eventually. So a team gathered to brainstorm ideas to get out of the situation. The solution they came up with: releasing an extended-length DVD of the movie.

The rationale was that an extended-length DVD would be hugely popular with *King Kong*'s biggest fans, and it also was projected to rake in profits that would cover enough of the shortfall at the box office. With these new projections, GE wouldn't have to lower the value of the franchise. It would avoid the need to take a charge and tell investors that its movie-making hadn't been as profitable as expected.

These sorts of tricks were legitimate under accounting rules, GE employees told themselves, and auditors, including the independent auditors from KPMG, had signed off on them. Auditors from the outside accounting firm worked in tandem with members of the Corporate Audit Staff to vet the handling of the company books. Finance workers at the company noticed that the KPMG staffers they worked with directly tended to be more sympathetic to GE's accounting decisions than their peers in the national office.

In 2007, other parts of GE's accounting caught the attention of a less sympathetic set of eyes. In January, the Boston office of the Securities and Exchange Commission informed GE that it was opening an investigation into the way GE was hedging the interest rate risk on the massive amounts of commercial paper it issued.

GE's practice of issuing short-term commercial paper to raise cash and lending it long-term at higher interest rates caused the source of funds and the subsequent loan to have mismatched durations. Commercial paper had to be repaid within nine months, but the resulting loan might last decades. Because of the mismatch in duration, the interest rate on the commercial paper could rise over time while the rate on the long-term loan was locked in. Changes in interest rates could squeeze GE profits, so it hedged that exposure by using derivatives. To avoid earnings volatility from the regular changes in value of those derivatives, the SEC allows companies to use specialized accounting for hedging. GE had fallen afoul of those rules and, according to the investigation, tried to cover it up to avoid taking a $200 million charge to its profits. KPMG auditors working at the company allegedly signed off on the inappropriate moves without approval from superiors in the national office.

August brought more heat as the SEC issued a formal order of investigation in the matter after spending months reviewing company documents and records. The potential charges were serious, especially in

light of the suspicions that GE's steady earnings were being carefully managed under the cover of its incomprehensible complexity. For some, the investigation was proof of misconduct that GE's size and reputation allowed to be easily swept under the rug.

With the SEC investigation, the wolf was now inside the door. As investigators dug in to scrutinize GE's accounting decisions, they didn't have to limit their scope to the commercial paper hedging issue. Their mandate allowed them to reach back into the past to see if the company's accounting over the years had been misleading. And the investigators were free to take their time doing it.

All GE could do was cooperate with information requests and answer questions. The investigative team soon found a slew of questionable practices that showed the scope of GE's ability to stretch, manipulate, and ignore accounting practices that are required of all public companies for the protection of investors.

The investigators found that if GE decided to sell an asset to boost quarterly profits and couldn't find a quick buyer, it sometimes called up someone close to home to buy the asset. Years earlier, in 2002 and 2003, GE had sold a number of diesel locomotives to banks, knowing that the banks would resell them months later. Six of these deals had been approved by internal auditors, even though GE maintained the risk related to owning the units and kept the engines running on GE property.

In a much bigger mess, investigators found fishy accounting on sales of jet engines and parts. GE had falsely boosted its profit margins on jet engine sales by combining low-margin engine sales with expected future sales of high-margin parts to come up with a higher "average" margin. It then used vague balance sheet entries so that the difference could be spread out over time.

GE's own accountants had grown worried that the accounting wasn't proper as far back as 1999, a finding confirmed by its auditors, and worried that fixing the books would result in taking a charge of $1 billion. Despite their concern, and without auditor objection, GE decided to continue accounting for the engines in the same way, while funding a write-off against the deferred balance so that it wouldn't continue increasing. In layman's terms, the company had set aside money to cover up the improprieties.

There was also a separate problem with accounting for engine parts:

an internal GE business learned that it was paying higher prices to another division for parts than some external customers. To ensure that it used the lower price in its accounting, the company created a workaround. Because the parts were part of a customer service agreement (CSA), the related revenue was recognized using a method based on the incurred cost in relation to the length of the agreement. If the prices were adjusted down to the more competitive rate, the total calculated cost of meeting the obligations of the contract would fall and cause an instant boost to revenue and profit called a "cumulative catchup." The move allowed GE to adjust for the newfound profitability of an existing contract, updating its value to reflect the asset's true worth.

It isn't hard to see why a business might want to do this, but a GE accountant shot down the idea with a stark reminder: it wasn't possible to increase the earnings of the entire company based on an internal price change.

A little imagination yielded another solution. By transferring the engine parts at cost (rather than via an internal sale) to the CSA business, the contract's profitability rose and triggered a contract accounting adjustment. The outcome was a large gain that offset the other accounting loss needed to eliminate the ongoing margin averaging in the engines business. The so-called catchup added about $1 billion to revenue and profit, thus offsetting an $844 million hit from removing the parts from the averaging scheme.

Naturally, this move wasn't allowed under accounting rules. It wasn't consistent with GE's other accounting, and companies are prohibited from making such discretionary changes.

And the $156 million difference that resulted from the change? GE also appeared to break the rules on that. Instead of recognizing the gain, the company took a page from the methods allegedly used during the Welch years: it banked it, in order to offset later earnings shortfalls from the internal price change.

Immelt's consistent declarations about the company's astute risk management and the dependability of his own management sharply contrasted with what the SEC was uncovering. Even some inside the company saw a culture that didn't match Immelt's spin that Capital was

superior than the banks with which it competed because it was imbued with the "magic of management" for which GE was known.

"Our financial services businesses are inherently more valuable than those of traditional banks or other financial services companies. Why? Because we have significant global origination in end-user markets that we understand better than others," he wrote to investors around this time.

These kinds of reassurances would take Immelt only so far as the timbers of the entire US financial system began to creak. In mid-2008, the country's largest mortgage lender, Countrywide Financial, had sold itself in a fire sale, shocking an already turbulent market. The Federal Reserve had slashed interest rates in the last months of 2007 and already cut them twice again in January 2008.

Capital had been a source of anxiety and worry for Jeff Immelt since his earliest days at the top of the company. He liked the people and loved the results they produced. He also enjoyed having the accompanying seat at the table with Wall Street power players. But in the end, Capital was always a problem. It was utterly complex and filled with risk, and its tentacles reached everywhere in the company. In addition, the broader market had lost its former patience for letting big industrial companies mess around in the risky business of finance.

Dabbling in financial services had virtually destroyed GE's historical rival Westinghouse. In some GE Capital offices, there were framed articles about Westinghouse's fall in the 1990s, a collapse linked to its overexposure to financial services and commercial real estate — along with a bad bet on its core business, the power generation market.

To achieve his visions for growth, Immelt felt that he had to optimize the GE portfolio for the long term. That vision would become his focus. Immelt's trust in his own ability to see strategic opportunities years in the distance was matched only by his trust in GE's operational rigor. His challenge was to set the company on the proper course to success and profit. The legendary management ethos baked into the DNA of the place would ensure that the right steps were taken to execute his plans. Jeff was the captain of the ship; his obligation was to stay at the helm, not get bogged down in the intricacies of the engine room.

This sort of gut-driven management style could be rationalized —

or explained at the very least — to the industrial divisions. But many of Immelt's grand visions fell flat at Capital, and that reception may have contributed to his distaste for that business. The marketing-based commercial strategy Immelt was imposing on GE's industrial units didn't really translate to financial services. And his optimism-driven management style, all pep talks and insistence that sheer determination to prevail could surmount any obstacle, didn't go over well with a division that measured success and failure in fractions of a percentage point and whose analyses almost always turned on quantitative factors. Capital's leaders knew what they were doing and understood the business, but they also faced enormous pressures, and certainly had many incentives, to make their targets at all costs.

Capital had a hard time trusting Immelt and his marketing push. Many felt that some of the declarations emerging from GE corporate headquarters reflected a fundamental lack of understanding of the financial service industry. To them, what worked in selling appliances wasn't going to work at Capital.

How well did Immelt understand the complexities of GE Capital and how the gains it earned and the risks it took helped or hurt the rest of the company? The question is unanswerable. Immelt's supporters scoffed at the notion that he failed to understand one of the most important profit drivers in the company. But some of his closest advisers soon began to question, in private, how well Immelt understood basic financial concepts. Their attempts to explain to the boss what they feared he didn't fully grasp were proving futile. He always waved away their doubts.

How much of Immelt's response to these efforts could be explained away as self-deprecation, especially on the part of an executive for whom it was a first instinct in a room full of strangers? Company flacks liked to remind newspaper reporters and others who challenged his decisions that Immelt had studied applied mathematics at Dartmouth. He'd gone on to the nation's preeminent business school — not exactly a friendly place if you don't like math. But Immelt also sometimes joked that his math skills were so poor that he couldn't help his daughter with her homework.

Then again, these were answers to the wrong question. Finance is numbers-driven, but it isn't just about math.

Jack Welch had often expressed amazement at Capital and the com-

parable ease with which it made a dollar by loaning one, rather than by designing and building and selling a machine. Even in his best-selling memoir, Welch had confessed to a lack of understanding of the intricacies of the business, even as he pushed to expand the finance business and its importance to the company at large. Welch had even commissioned his staff to write a booklet of finance basics; he wanted to make sure he didn't get it wrong.

Some of his colleagues suspected that Immelt's background in sales and the industrial units had left him with a fuzzy grasp of the financial basics. His positions in those businesses had been meaningful, and he had been responsible for their finances. But GE's matrix management structure had assigned others to be the crucial arbiters on some of the most complex financial decisions in which he had been involved.

Financial services was different from anything Immelt had worked with up close. Making money from money seemed shockingly simple to him at first, as it had to Welch, but the balance sheets were treacherously complex, and deep risks lurked there and were not always easily spotted in the quarterly profits and losses. It was reasonable to think that Immelt didn't know much about how the greasy gears of Capital worked. In reality, it isn't clear that many people anywhere understood those complexities.

Still, pinning down how well the company's leader understood its crucial lending unit was impossible even for some of those who were closest to him. Some of Immelt's critics acknowledge that he was sharper on finance than they expected. But another who worked closely with him for years said that Immelt struggled with basic concepts — the difference between secured and unsecured debt, for instance, which was fundamental to a lending operation like GE Capital.

No one communicated any of these doubts, however, outside of GE. As 2007 drew to a close, Immelt brandished his usual bright outlook when talking about GE Capital. Immelt's optimism was a fortress.

"Our financial services businesses should do well in a year like 2008," he wrote. "There could be $300 billion of assets available at high returns. We plan to seize opportunities in the current turmoil and position our financial services businesses for years of profitable growth."

15

SIZE DOESN'T MATTER

IN SESSIONS IN the Crotonville pit for the newest GE hires, Jeff Immelt openly talked about his doubts and concerns about failure. But to the public, he showed no weakness, especially not when talking about his company or its future.

That style was in keeping with Immelt's leadership overall. He was dedicated to the legacy of Crotonville and GE's management ethos generally; he was also more committed to principles of diversity than some members of his board. He was certainly more committed to those principles than many had been in the Welch years.

Immelt won plaudits, even among those who thought him a less than brilliant businessman, for a genuine commitment to improving GE's mediocre track record on diversity in race and gender. The company fame wall at 30 Rockefeller Center looked like an early twentieth-century snapshot of the US Senate: stern, stiff, bone white, and male. And the company's gender politics, even among its defenders, was dubious.

Take Bill Lane, Welch's speechwriter, still warmly remembered by some of the company flacks. His memoir of his GE years provides a refreshingly candid glimpse of Welch and the company's foibles, as well as the tone-deaf chauvinism of General Electric in the depths of the 1980s and '90s. In one early passage, Lane gleefully relates an episode from one of his first retreats at Crotonville: one evening a beer-fueled posse of attendees found a way to project rented porn videos on the big screen of The Pit, the leadership training auditorium. "The two 'girls' in our class,

almost as wild as the worst of us, shook their heads and took a pass on this particular evening's entertainment," Lane reports.

His Boswell reports Jack Welch's reaction: "He laughed and purred, because he perceived, correctly, that the juvenile, *Animal House* behavior was a welcome symptom of the high-spirited *joie de vivre* that was spreading across his company. Our behavior reminded him of our early days in the plastics business — and that was just fine with him."

Whether that *joie de vivre* extended to the women of the company is not explored in Lane's anecdote. Other than escaping from the shenanigans that evening, the reactions of their female colleagues to GE's men merit no mention in Lane's account of Crotonville and the bonding that happened there.

Immelt, however, was determined to change the boys' club reputation of GE and to affirm its rightful inclusion among the corporate giants of the twenty-first century. He would keep the vital bits of the Welch-era GE and leave the bad boy stuff behind.

Immelt projected a bold and expansive year for GE's financial services business for 2008. He acknowledged the uncertainty and turmoil in the markets, but predictably added that any weakness would bring an opportunity to buy up cheap assets. As Immelt described the tumult: "The financial market risk is repricing."

In reality, GE was pumping the brakes hard. Mortgage delinquencies across the United States had been rising well above normal through 2007, and those delinquencies were highest among those with poor credit. The major banks had begun to mark down the value of their mortgage portfolios by billions of dollars, causing some turbulence in the commercial paper market, where many borrowers were using mortgage-backed bonds as collateral.

GE Capital saw that a storm was coming in the mortgage market, especially in the type of low-quality loans arranged by the WMC unit. GE's mortgage bet looked badly timed. In addition, it may have picked the worst possible partner to anchor the strategy. The group originated more than $65 billion in mortgages from 2005 to 2007, making it one of the top producers of subprime mortgages in the United States.

Like many companies originating loans, GE's WMC mortgage unit would typically sell the mortgages to investment banks, which then re-

packaged them into bonds that were repackaged again for investors. The bonds weren't GE's, but GE did create the underlying loans, many of which weren't as they seemed. Lending in the WMC division was running wild as workers hustled to close deals, even those with mortgage applications that included what they knew to be false information. Some rejected applicants found a path to approval by reapplying weeks later and reporting drastically higher income or by clearly misrepresenting other information on their application.

One applicant's occupation had been reported as "museum curator," but it turned out that the applicant actually worked in a hair salon. Another reported being an ice cream truck driver with an income of more than $5,000 a month and used customer letters as proof of self-employment. But even when such bogus applications were flagged to supervisors, they were still approved.

Every appearance of blatant lender fraud was there for the observing. Soon, when banks started refusing to buy certain loans from WMC because of problems or suspected fraud, WMC began shipping the same loans to other banks without disclosing the reasons for the previous rejections, or even that they had been rejected. Some banks were forcing WMC to repurchase the loans, which it then held on its own balance sheet. WMC reviewed a sampling of almost 1,300 loans it bought back in 2005 and found that the files on 78 percent of them "contained at least one piece of false information."

This behavior was encouraged by the pressure at GE to meet profit targets, which required more and more loan volume. GE's own internal auditing team warned in April 2007 that the business had "jacked up volume without controls."

The bosses at General Electric didn't waste time running for the door when they saw the fire spreading. GE stopped WMC from writing any more loans, laid off more than four thousand people, and sold off $3.7 billion of the assets. In 2007 alone, WMC had racked up about $1 billion in losses.

Still, GE executives congratulated themselves for their success in exiting such a turbulent market. But analysts were confused by GE cutting and running on the mortgage business entirely. The cowboy ethic of GE Capital usually prized running toward a crisis to double down

and scoop up cheap assets. Immelt left no doubt, however, that GE just wasn't interested in being in the mortgage business anymore.

"I just think we wanted to get this off the table vis-à-vis the things that investors have to think about with GE," he said. Keith Sherin assured investors that the financial services division was cutting costs and lowering the risk in its portfolio.

With doubt growing over housing market exposure everywhere, GE made a show of strength. The company supersized its ever-present stock buyback program and bought $14 billion in stock. The subsequent dividend payment would shovel over $26 billion in cash to shareholders.

For Jeff Immelt, WMC was just more proof that GE Capital was an unending puzzle filled with risk. He knew that GE needed it, but it wasn't clear how to separate, shrink, or somehow change Capital into something else. In the meantime, he followed his predecessor's lead by happily using the division as a tool for producing profits.

Jack Welch is usually blamed for unleashing the beast of GE Capital, and it certainly did grow quite large under him. It was also Welch who had made Capital into a business that was the organizational equal of any other part of GE. Welch claimed that he always held Capital at bay enough to appease the credit rating agencies. To Welch, the most important GE weapon was its triple-A credit rating, which allowed for cheap borrowing.

But Welch also had received regular Capital headaches. His foray into the brokerage business — acquiring control of Kidder Peabody in the 1980s — was a disaster. There was an insider trading scandal, years of losses, and finally a massive bond-trading debacle that led to the sale of most of the firm to Paine Webber in the mid-1990s. GE ultimately took a $1.2 billion loss on the whole episode.

Welch was aware of, and seemingly self-conscious about, the role that Capital was playing. He famously protested to *Fortune*'s Carol Loomis in 1999 when GE's classification in the Fortune 500 was moved from "electrical equipment" to "diversified financial services." If GE was going to look so much like a bank, *Fortune* would classify it as one. Loomis later reflected that Welch was probably mostly upset about the multiple: Wall Street valued the shares of financial services companies at a lower rate than those of cash-rich industrial companies.

Now, under Immelt, there was still a gaping separation between the industrial business and GE Capital, one that could be characterized as a schism between the primarily blue-collar GE workforce and the buttoned-up office workers of Capital. Those at Capital viewed themselves as the ones printing the money for GE while the industrial businesses struggled to achieve growth or make beneficial acquisitions. To the industrial teams, Capital didn't produce anything tangible and its value came only from financing sales of industrial products.

In reality, GE Capital was able to attract some of the best people in the finance world. They cultivated the wizards of finance who didn't want to be on Wall Street but were attracted to working at one of the world's biggest and best multinational corporations while essentially operating a high-powered hedge fund.

This reputation and the timeless respect for the brand were essential to GE Capital's success. But such magnetism is hard to replace once lost.

The rank and file of Capital, if they stayed at GE, generally stayed at Capital. GE bankers closing billion-dollar deals while working twenty minutes away from the country's wealthiest towns generally weren't yearning to move their families to Schenectady, New York, or Greenville, South Carolina, where GE had outposts. And when workers across GE were pushed to get their Six Sigma green belt training, a defect reduction system usually used in manufacturing, many GE Capital workers resisted, seeing the training as more management mumbo jumbo that just didn't apply to finance. So, while GE Capital was expected to be in lockstep with the rest of the company, its autonomy and distinct culture also contributed to the schism between the financial services division and the industrial divisions.

In the eyes of Capital workers, the role of the sales organization shifted when Immelt became CEO. Salespeople could push around finance people, but they weren't held accountable for outcomes. If a deal closed, it was celebrated as a success of the salespeople. If it fell apart, the blame seemed to fall on the risk and finance team. The culture created by this dynamic led to executive promotions of people who were underqualified for complex finance jobs. Some were not only unprepared for their roles but also unfamiliar with basic finance concepts, according to people inside the business at the time.

During a weekly credit meeting, a new senior manager asked about EBITDA, the finance acronym for "earnings before interest, tax, depreciation, and amortization." Removing those items from the analysis provides a better comparison of two companies, and the resulting figure can be a proxy for cash flow. The executive wasn't raising concerns about the group's estimates or calculations. Rather, he'd never heard the term and just wanted to know what it was.

Many found that level of ignorance shocking and dangerous. After all, this was GE Capital, where they prided themselves on intelligence and expertise. Recruiters, sensing the friction this situation created, began picking off Capital talent throughout the ranks, and often people with less experience were hired to replace those who left.

The complexity of GE Capital's businesses required sharp people. Some seemingly reckless moves caused concern among those in the division. At one point, an executive based in the Midwest was running a senior risk management function that was based in Norwalk, primarily by phone.

Now, as the mortgage market slowed and markets began to froth, teams still had to make their financial targets. When business was slow, Capital workers got creative. They would try to edge into their colleagues' regions or business areas, looking for deals. Old contracts were opened for renegotiation to pad the deal count for a quarter.

The pressure made for a more stressful work environment. Verbal abuse from colleagues or superiors aimed at both sexes, sometimes in front of large groups of other employees, wasn't unusual. Women in the division complained that they were less than respected. Anyone who was different could feel pressure to conform — even growing a beard sometimes drew the jockish ire of colleagues. GE's Human Resources Department allegedly had to send monitors into the Capital offices occasionally to make sure everyone was behaving.

Even while de-risking, GE Capital was still placing bets. When it bought Poland's Bank BPH, GE Capital became the owner of one of the biggest banks in that country. The bank's signs were changed to highlight the new ownership, which now held billions of dollars in Polish mortgages, many with adjustable rates, denominated in Swiss francs.

GE was signing up aggressive deals all over. In 2007, it acquired the

aerospace business of the Smiths Group for $4.8 billion, the oil and gas operations of Vetco Gray for $1.9 billion, and a diagnostics business from Abbott Laboratories for $8.1 billion.

Immelt and GE Capital also dove into commercial real estate, chasing a hot sector. Exposure to real estate had been rising in Immelt's early years, but he accelerated in 2006. Through a number of large acquisitions of real estate investment trusts and then loads of smaller deals, the company built a huge portfolio.

Immelt was starting from a huge base. In 2000, the company already had real estate assets of more than $20 billion. Cutting a deal in December 2005 to buy Arden Realty for $3.2 billion, one of the company's biggest-ever real estate purchases, brought GE's total real estate assets to almost $50 billion. That rose to almost $80 billion by the end of 2007, with about half in the form of equity ownership in properties.

The buying continued. When GE bought a real estate investment trust, Crow Holding's third fund, for $2.2 billion, it acquired industrial buildings, retail shopping centers, and hotels. Dundee REIT, bought for $2.3 billion, contributed commercial properties in Canada, while other deals brought holdings in Sweden and shopping centers in Madrid.

GE Capital pushed hard in Japan in 2007 as GE tried to more than double its holdings while commercial properties were soaring in value. The head of the real estate business in Japan, Tomoyuki Yoshida, outlined the strategy to *Bloomberg News,* giving insight into the orders from the top. "Even with land prices increasing so dramatically, Japan still looks attractive to us," he said, adding that GE had "no ceiling" in its approach.

"Our main target could be companies that own lots of real estate. Size doesn't matter to GE."

16

THE SCREWUP

JOHN FLANNERY WAS in Tokyo playing a familiar role: managing a business and then cleaning up a mess. He had moved to Japan in 2005 with his family to run GE Capital's Asia-Pacific operations, after spending most of his life in Connecticut.

The typical GE executive was an outgoing, social creature, seemingly born with a list of talking points memorized and a PowerPoint slide clicker in his or her hand. In contrast, John Flannery could sometimes come off as a little bit shy. Offstage, he was wry, shrewd, and friendly, but the spotlight onstage wasn't his favorite place. He was more at home working the numbers behind the scenes.

Flannery's résumé couldn't have been more focused on banking and corporate finance. After leaving Wharton, he hopscotched around GE Capital: three years in the leveraged buyout division, four years in the corporate restructuring group, three years in the equity capital group, a two-year stint in Buenos Aires running Capital in the region, three years overseeing the media and consumer group in the equity division, a year overseeing the whole equity division, and two years overseeing the bank loan group.

A number-cruncher by nature and self-declared "foodie," he tore through a wide variety of books, sometimes simultaneously reading a nonfiction work and a novel. The relocation to Japan was a new challenge and the next step in his rise at GE. But it wouldn't exactly be a vacation. Flannery had his work cut out for him.

Upon moving overseas and touring the GE operations, he started to get a better view of the frustrating bureaucracy that permeated every piece of GE. A relatively simple issue like the difference between time zones compounded the stress of getting all the bureaucratic approvals he needed to win back in Connecticut. GE's bureaucratic structure made it dramatically harder to keep the overseas operation running and the company's profits growing. Flannery thought that decision-making power needed to be more decentralized, so he could make more of his own decisions, rather than waiting for the plodding corporate Borg back home.

At the same time, as others at Capital were seeking ways to curtail risk in the portfolio, he had to find a way to deal with one of his most prominent businesses blowing up. Acquired a decade earlier, the consumer finance business called Lake had suddenly become a headache in 2006, when a change in Japan's consumer lending laws sent the once-lucrative market into chaos. Lake had been a cash machine for GE, and investors loved it. The unit borrowed at very low interest rates and lent out funds at much higher ones — easy money.

But public pressure had persuaded Japanese lawmakers to change usury laws, capping the top interest rates on consumer loans at 15 to 20 percent, down from close to 30 percent before. Compounding the impact of the new law was a ruling from the high court of the country that existing loans with rates above 20 percent were illegal. That cleared the path for some of Lake's customers to demand refunds for excess interest payments, an exposure that prompted banks to increase their cash reserves.

Overnight the Lake business model had gone from inescapably profitable, if morally dubious, to illegal. This was now John Flannery's problem to solve.

The market had turned hostile on GE, and potentially large future liabilities loomed from the growing threat that the company would have to refund customers. Flannery, who had done hundreds of deals in his tenure at Capital, quickly began to take action. He made plans to close nearly two-thirds of Lake's branches. Citigroup, another lender caught in the same pickle, had gone even further, essentially shuttering its entire operation as losses piled up. GE also shifted the Lake unit into what it called "discontinued operations," a convenient bucket of accounting

that enabled GE to exclude losses from the adjusted profits the company highlighted to investors.

Meanwhile, a scrambling Flannery found a savior: he cut a deal to sell the Lake business to Shinsei Bank for $5.4 billion, along with a loss-sharing agreement to help the new owner split the cost of the coming wave of refund requests. For GE and Flannery, escaping from the Lake fiasco with a $1 billion charge to earnings was cause to declare victory. It could have been much worse.

For Flannery, it was also a lesson in the difficulty of managing a company seven thousand miles away from headquarters on the other side of the planet. Even as he scrambled to quickly limit the company's exposure, GE's management and operational structures, exacerbated by the distance, had led to painful delays in getting approvals and answers back from Connecticut. Flannery could see that GE's approach to running its international operations was deeply flawed.

Meanwhile, back in Connecticut, Immelt's secret weapon was back by his side, and just in the nick of time. Beth Comstock — the pistol of a chief marketing officer who had snuck away for more than two years to run NBC Universal's digital operations — came back into the fold at GE and into her previous job.

Comstock had struggled in the last assignment, where her signature move had been the $600 million acquisition of ivillage.com, a women-oriented content site. That purchase had been a kind of land grab in the digital world, but the price tag was openly mocked, partly because NBC spent more buying ivillage.com than News Corp paid in its infamous purchase of the trendy social media site Myspace.

In addition, the cultures of the companies didn't match, and Comstock's idea to launch a TV show connected to the ivillage.com brand also fizzled. Notably, following Comstock's return to the conglomerate's mothership after the disastrous deal, NBC broke up the broad responsibilities of her position among existing executives.

The stumble didn't stunt her career. Long an Immelt favorite, she was back controlling messaging and overseeing the never-ending marketing push for GE. Then a major crack appeared inside GE Capital in early 2008.

In mid-March, Immelt assured investors that, despite the brewing

storm in financial services around the world, the company's first-quarter performance was on track and GE would be fine. The underlying message was that the conglomerate model — divisions picking up the slack when other divisions were hurting — was still working well.

A few weeks later, GE's first-quarter results hit the newswires at 6:00 a.m. The company had missed its targets. Badly. Some thought that the reports were somehow mistaken — but they weren't.

GE Capital had taken a massive hit from the ongoing mortgage-related disruption. GE had boasted that it knew the markets better than others and would sail through any trouble because of its risk management mastery. Now it was in the same boat as every other financial firm — explaining to investors that the collapse of Bear Stearns had hit harder than expected. The GE earnings engine had thrown a rod.

GE had prided itself on always making its numbers, squeezing out what earnings it needed, even in a flurry of last-minute deal-making, to deliver to investors exactly what had been promised, or even a little more. And that streak had largely been the work of GE Capital, which had always been there to locate the necessary pennies per share, always just in time. The first quarter of 2008 showed once and for all that it had long been GE Capital that was the force keeping the earnings streak on track — and that GE's earnings were unpredictable when Capital itself was struggling.

The market turmoil had inched itself closer to General Electric. Keith Sherin said that the meltdown around Bear Stearns had created more volatility in the capital markets "than we could have ever anticipated." But to GE investors — to those who truly believed in the conglomerate model — that answer didn't make sense.

Immelt was blunter. The commercial finance division experienced an "inability to do transactions in the last two weeks that we normally could get done" and the revaluations on some assets were negative. In other words, those crucial last-minute deals that GE Capital typically pulled off — selling assets and making other adjustments — didn't happen.

Seeing the warning sign, GE shareholders started to sell. The stock saw its worst fall in years on the news and lost almost $50 billion in market value that day.

Adding to the pain, as Wall Street still digested the news that week,

Jack Welch went on CNBC, a network GE owned, and took the not-so-subtle path of ripping Immelt apart on live television. If investor confidence had been teetering, Welch pushed it off the cliff.

"Here's the screwup," said Welch, talking about Immelt. "You made a promise that you'd deliver this and you missed three weeks later. Jeff has a credibility issue. He's getting his ass kicked."

Welch's orders were clear: Don't do it again. Do what it takes to keep the steady earnings flowing.

"I'd be shocked beyond belief and I'd get a gun out and shoot him if he doesn't make what he promised now," he said. "Just deliver the earnings. Tell them you're going to grow 12 percent and deliver 12 percent."

For those who didn't think Jeff Immelt was making the right decisions, they now had their infallible CEO idol making the very same point. Immelt didn't respond publicly to the lashing, and Welch eventually tried to walk back his criticism. But the horse was long out of the barn.

Welch's patriarchal dressing-down scorched Immelt. The message, heard all the way down to the factory workers, was that Immelt made an egregious error and violated GE's sacred devotion to delivering consistency to investors. Suddenly Immelt, who had cultivated, with boatloads of media fertilizer, the aura of having all the answers to the toughest questions in business, didn't seem so omnipotent after all.

On Wall Street, the earnings miss and Welch's comments started chatter of the need to break up General Electric. That had been an almost unthinkable option under Welch. But after decades of veneration, the underlying logic of GE's structure was now being questioned. In a sense, the street was belatedly noticing what many in the business world already knew: that devotion to its model had permitted GE to operate in an ungainly conglomerate structure that had long fallen out of fashion for everyone else. GE had gotten away with it because, well, it was GE, and the numbers were good. But now that the numbers weren't good, investors started to question whether simply being GE was still good enough.

To Immelt's chagrin, these questions were reflected in his stock price. The price of a share of General Electric continued to fall. Within a few weeks, the price had fallen below $30.

ENTER THE BEAR

L IKE ANY MODERN bank, JPMorgan Chase had to juggle its interests. All the parts of its sprawling operations, from retiree savings accounts to complex derivatives, were covered by regulations designed to prevent conflicts of interest or self-dealing at the cost of clients.

Steve Tusa was a sell-side analyst in the bank's New York headquarters, and he was a good one. As a sell-side guy, his job was to rake through the financial records and public utterances of the public companies he covered and find unappreciated problems before they hammered the stock price so that every investor who paid for his research would have a chance to escape to safety.

Now Tusa had found some problems in the financial disclosures of one of the bank's biggest and most historic clients: General Electric.

GE wasn't just a hugely important company, with a combative investor relations department that didn't take kindly to negative research notes from Wall Street analysts. It was also an enormously important customer of JPMorgan Chase, which had become a go-to banker for GE over the years.

The inherent conflicts of interest embedded in the Wall Street research models employed by massive banks had supposedly been corrected by a $1.4 billon settlement with regulators in 2003. Regulators had come down on the banks in an effort to erase the pressure on their analysts to publish overly positive research on companies that would

help their firm's bankers land lucrative investment banking deals. It was deals and trading commissions, not research, that made money. But even years later, the stock recommendations still skewed strongly to the positive as companies found other ways to incentivize analyst behavior.

Sophisticated investors with their own internal analyst teams had eventually found the watered-down research from banks to be of little use. Instead, the real use for analysts was access. Investors depended on their analysts to arrange private meetings with top-level executives at companies. Private meetings, however, also created pressure to come up with positive ratings. Even openly negative analysts admitted how difficult it was to assign a sell rating on a company that could suddenly stop providing valuable one-on-one time with executives. They couldn't afford to get cut off.

Regardless of the relationships within JPMorgan, Tusa's doubts were real. He had been on the team covering GE for years, and a couple of years earlier he had fired what amounted to a warning shot across GE's bow. In the spring of 2006, in a research note, Tusa had noted that GE Capital's tax rate had come in unexpectedly and conveniently low — 14 percent compared to the 17 to 19 percent that the company had previously projected. A sharply lower effective tax rate had the effect it would have on any household: for that quarter, GE recorded that it had taken home more earnings. That was the sort of thing that investors loved. It was also what analysts like Tusa called "low-quality earnings" — a slightly more polite way of saying "awfully convenient."

This was not an instance of profit from additional goods sold or loans closed. And as with any tax rate–generated gain, there was no guarantee that the same benefit would be there just three months later, in the next quarter. Calling out low-quality earnings, as Tusa did, was coded Wall Street jargon that signaled to sophisticated investors, and to the well-heeled, that GE's public declarations about how well the business was doing weren't quite what they seemed.

JPMorgan's relationship with GE went all the way to the top. As Immelt was buying and selling business units at a relentless pace, he often turned to JPMorgan for advice. The dance partners had a relationship that went way back: for GE, the titular financier J. P. Morgan was arguably a much more important historic figure than Thomas Edison.

More important, the most eager chatterbox in Midtown, the irrepressible JPMorgan banker James Bainbridge Lee Jr., was working his way ever deeper into the inner councils of GE.

Jimmy Lee had a rapport, to hear him tell it, with virtually everyone in American business, but he especially loved GE. He had bonded deeply with Welch, known Immelt for years, and helped bring numerous deals from the big industrial at 30 Rock a few blocks over to JPMorgan Chase's glass-fronted tower near Grand Central.

"This doesn't sound good when I say it this way," he once told us, "but I'm like the only guy on Wall Street that knows them this well."

Lee's comments privately demonstrated the stakes for Tusa as he highlighted the rising questions around GE. A tough note from a Morgan analyst could lead to pressure in Lee's crucial banking relationship with a client like GE. But the fact that JPMorgan had a strong banking connection to the company also lent strong credibility to Tusa's more negative observations. To those who knew the research game, Tusa was the last guy who should rock a boat as big as GE. When he did, the thinking went, he must have had a good reason.

Tusa was a son of Greenwich, Connecticut, the Napa Valley of hedge funds and the bedroom community of Manhattan finance's elite. His father was a successful lawyer, and as a child he rode bikes with friends named Ian and Shep Murray, who would go on to found the uber-preppy clothier Vineyard Vines.

Tusa had earned a political science degree but skipped the standard path to finance of getting an MBA. Instead, he had headed straight into banking, and what he learned there on the job helped him, he believed, to understand GE.

He was also tough. He had a contrarian impulse and was more likely to cheer for a team to lose than to pick a winner. The analyst liked to say that his first career choice would have been playing center for the New York Rangers. Tusa played hockey competitively into adulthood and carried himself with an athlete's wiry intensity. When he first met one of us, Tusa was sporting a mullet, which he had grown out of superstition because his beloved Rangers had made it into the NHL playoffs.

As he sat in his Midtown office in 2008, reading yet another round of GE financials also gave Tusa a bad feeling of déjà vu from those early days at Morgan. He had arrived on the industrials beat just in time to

see up close the enormous collapse of Tyco International, a seemingly impregnable industrial company, known for its smooth and consistent earnings that fell to pieces overnight. Tyco had been a massive accounting fraud, but analysts missed the story. The smoothness of the earnings had rocked them to sleep long enough for the investors they served to lose everything.

The scandal had left a deep impression on Tusa, who emerged from it determined to be skeptical of everything. Though he maintained the peculiar faux-civility of the earnings calls, Tusa, like many of his colleagues, tried to ask tough questions without rocking the boat hard enough to get pushback from a company's investor relations department.

Now, as the head of the JPMorgan team that covered GE, he was resolved to dig deeper. Tusa had placed an "overweight" rating on the stock, which meant that investors should hold more of it than the average investment. An overweight rating was a tacit endorsement to buy the shares at their current price. Regardless of his published rating, however, Tusa's skeptical nature led him to keep a close eye on Immelt as he struggled to find growth and adjust to the new world of running GE without the supports that Welch had enjoyed.

When GE missed its earnings target, Tusa decided that the time had come. Dropping a bomb that shocked investors, the analyst downgraded GE from "overweight" to "neutral." This financial jargon might have held little meaning for an outsider, but insiders knew that it represented the withdrawal of an endorsement worth tens of millions at least.

Tusa explained himself. Speaking of GE in a note he distributed to his clients, he wrote: "It would appear as though accountability for hitting targets is the top priority, and some managers might be chasing earnings." He added: "We also think the high bar for success in such a competitive environment could create a scenario in which bad news is not tolerated, making necessary communication with senior level managers a challenge until it's too late to fix."

In other words, the bosses demanded results at any cost and problems could be hidden for the sake of preserving performance, thus allowing small problems to become big problems before they were detected.

18

PAPER JAM

BEVERLY HILLS WAS mild and clear on the evening of September 11, 2008. The last person you expected to see among the swaying palms of Los Angeles on that particular night was Jeff Immelt.

On the other side of the country, huddled in the deep canyons of downtown Manhattan, the most powerful minds on Wall Street were wrestling with a financial contagion that had choked the economy and would topple some of the world's mightiest financial companies.

General Electric wasn't technically a bank, but its finance organization dwarfed most banks in the country. Because of its industrial core and positive messaging to the market about its condition, not to mention its reputation, the company wasn't on anybody's radar as the crisis deepened.

But that ended quickly.

A few days earlier, Immelt had made a call to Treasury Secretary Hank Paulson to tell him that GE was having some problems selling its commercial paper to investors. (As noted earlier, the paper is short-term debt that companies use to meet the cash needs of their operations.) Commercial paper was GE's lifeblood. It used its stellar triple-A credit rating to get cheap access to tens of billions of dollars in commercial paper to keep GE Capital running smoothly. If GE couldn't access that market, it couldn't meet its massive obligations. It would be technically insolvent.

Paulson was alarmed by the call, he later wrote in his memoir *On the*

Brink. He and Immelt had a history together. They did business when Paulson was a banker and executive at Goldman Sachs. Both had played football at Dartmouth, although they had graduated a decade apart, and both attended Harvard Business School. Paulson was known to be dead-serious, while Immelt always seemed to have time for a joke.

These were not ordinary times. Just days earlier, the US government had seized control of mortgage giants Fannie Mae and Freddie Mac, and things were about to get a lot worse. For GE, the storm was in sight.

In Beverly Hills, Immelt was heading to the Hotel Beverly Terrace and a restaurant called Trattoria Amici. He wasn't meeting government officials, bankers, or advisers to help get GE through the financial apocalypse. Rather, he was meeting with Steven Spielberg, along with Universal Studios president Ron Meyer and DreamWorks CEO Stacey Snider.

It was purely social, the companies assured the show business press, not a deal meeting.

Immelt, like his predecessor, enjoyed being involved in the media business. Owning NBC, and eventually Universal as well, had brought the industrial giant onto the red carpets of Hollywood and into the halls of Rockefeller Center, recalling the days of giant corporate conglomerates that could, and did, own anything.

After GE bought Universal Studios, Immelt used to show off previews of blockbuster movies coming out of the division at the large executive leadership meetings of internal managers. The workers in the crowd, who were from all divisions, weren't always impressed with the highlighted big-budget output of the movie studio.

During a Q&A session in one meeting, an employee asked Immelt how it felt to be a media mogul. The boss gave a characteristic full laugh and snarky response.

"I'm not a media mogul, I'm the media mogul's boss," he told the group.

Even in the depths of the financial crisis, Immelt had joked to people that "the good news is that we own CNBC," which saw skyrocketing ratings during the never-ending news cycle.

Meanwhile, back in New York, regulators and the heads of the major banks were facing an inconceivable weekend of round-the-clock work to avoid catastrophe in the banking system. By Monday morning, Merrill Lynch was sold and Lehman Brothers had declared bankruptcy.

He might have been in Beverly Hills, but Immelt was closely following GE's exposure and navigation during the crisis. He had flagged the difficulties that GE was having in the commercial paper market to regulators, but he hadn't communicated those concerns to investors or the public. In fact, GE was explicitly communicating the opposite by insisting that it wasn't having problems. The coming weeks would bring recurring assurances from GE that everything was fine, but the wide gap between those assurances and reality couldn't remain open for very long.

In fact, precisely what PIMCO's Bill Gross had warned about in 2002 was now coming to a head. He had said that GE's dependence on commercial paper was dangerous because the company wasn't fully backing up its needs with a contingency plan, as many financial services funds did, or were required to do.

Trust and confidence are everything in finance. People put their money in a bank only because they are very confident that they can get it back if needed. If their confidence wavers, they'll pull their money out and stop using the bank. When that happens all at once, the bank fails. For its part, GE certainly didn't want its counterparties to get spooked; that could increase the cost of borrowing or even freeze the market altogether.

On September 14, a Sunday, GE's investor relations department issued a letter reiterating that the company was in healthy condition and maintaining that its commercial paper programs continued to be "robust" and that "we are not raising external capital and have no need to."

But the very next day, Immelt showed up at Paulson's office in the evening to again sound the alarm about potential problems at GE. He told the treasury secretary that the commercial paper operation was getting worse and that GE was having a hard time selling debt that lasted longer than overnight.

That Monday night, New York Federal Reserve president Tim Geithner had a meeting on his schedule for 7:00 p.m. with a simple title: "GE Issues."

The financial crisis changed the world's view of commercial paper. Although it had long been considered easy to access, fully trustworthy, and a liquid source of funding, the crisis raised doubts and caused much of the commercial paper market to disappear. This is a real problem when an entire portion of the economy is depending on it to function.

The collapse of commercial paper was similar to a gasoline crisis. Roads and highways were built with the assumption that gas will always be available. So if all the gas stations go empty, the entire automobile transportation system eventually stops working. And when the gas stops flowing, the reaction is predictable: people panic, prices rise, and some start hoarding, but there's little that most people can do.

Commercial paper was not a new concept. It was how Marcus Goldman and Jeffrey Sachs got their start on Wall Street after the Civil War. As a tool that allowed companies to quickly borrow or lend, preventing costly cash shortages and bringing them a return when there was extra cash, commercial paper met a need to stabilize the cash holdings of businesses.

In their research, two finance professors from New York University's Stern School of Business found that there was about $2 trillion in commercial paper outstanding in the United States at the beginning of 2007. The first disruption began just months later, in late July, when two Bear Stearns hedge funds went bankrupt because of massive losses on subprime mortgages, causing other funds to reassess their own holdings in similar mortgages and mark them down. The moves triggered a cascade of fear and devaluations. Some funds halted client withdrawals because they couldn't value assets fast enough.

The mortgage contagion infected other businesses as investors suddenly became cautious about using commercial paper, a large portion of which used mortgages as collateral. This set the stage for another shock to the market a year later when Lehman Brothers went bankrupt over a weekend. Investors holding commercial paper sold by Lehman found themselves without a chair when the music stopped.

The most notable impact of Lehman's failure was the major losses taken by a money market fund, the Reserve Primary Fund, with $65 billion in assets, when its shares fell below $1, breaking yet another source of a once-dependable investment. People rushed to salvage anything they could by trying to get their money back — it was a full-fledged run on the bank that stopped on September 19 only because the US government jumped into the fray and said that it would insure investments in such funds.

This fixed the problem — but you can't unring a bell. If major sellers of commercial paper could go bankrupt overnight, that changed every

investor's assessment of the risks involved. When money market funds began deemphasizing commercial paper in their holdings, demand dropped as supply remained unchanged. It became harder and more expensive to access the market. For a massive user of commercial paper like General Electric, this was suddenly an existential problem.

On the morning of September 20, a Saturday, Jeff Immelt and Keith Sherin traveled to GE's Fairfield headquarters to meet with two bankers coming up from Manhattan. David Solomon and John Weinberg, top advisers from Goldman Sachs, arrived expecting a meeting about the crisis at hand and a discussion of potential defensive moves for GE. As the meeting got under way, it was clear that Immelt had other concerns.

Was Goldman Sachs going out of business and what would that mean for GE? Five days after Lehman's shocking bankruptcy, Immelt was worried that a Goldman collapse would destroy all remaining confidence in the markets. For some, if a bank like Goldman went down, there was little chance that GE could stay above the fray. The bankers left Immelt with a strong assurance that Goldman wasn't going anywhere.

Five days later, on September 25, GE again tried to reassure investors that it was performing well in a tough environment, but it was also clearly starting to show signs of stress.

GE lowered its financial projections for the soon-to-end third quarter to reflect "the unprecedented weakness and unpredictability in the financial services markets," which it didn't expect to improve anytime soon. GE again backed its commitment to its triple-A debt rating, an often-stated source of pride for Immelt, and said that it wouldn't need to raise additional long-term debt for the rest of the year.

But GE also ominously showed that it was moving to retain cash by suspending its stock buyback program, on which it had spent $278 million in September — even as it faced an increasingly difficult time funding its operations. GE's required regulatory filing on the news was dry, saying that the company would reduce its commercial paper levels. But its messaging to the public was more upbeat.

In referring to the commercial paper reduction, GE's press release said that it was making the moves "although demand remains strong" for its debt. Both Immelt and Sherin made the point that retaining more cash, especially in GE Capital, was not only prudent, given the ongoing uncertainty, but would also make it much easier to "participate in

M&A [mergers and acquisitions] that we plan on doing over the next 18 months."

It was dubious at best that GE might be considering buying companies while the sky was falling.

The marketing message was that GE was performing well compared to its peers, and that it was taking the precautions that would be expected of a highly regarded triple-A-rated company. When asked, Immelt ruled out raising equity capital by selling shares to the market.

"Again, we feel very secure about how the funding looks and the strength of the company and the strength of the balance sheet," Immelt told investors. "We really believe in our business model and feel secure that we're well positioned here."

But the spin wasn't working. The market needed a real sign of confidence in GE.

The company spent a lot of time and money on convincing the general public of its importance and central role in American life. Wall Street was indifferent to that and only worried about GE Capital, the financial services leviathan that produced half of GE's profits.

The continued caution around commercial paper was causing real problems for GE. In the market, spooked investors didn't want to hold the paper for too long, so it was being sold for shorter durations. As a result, companies had to sell paper more often, causing the number of issuances to skyrocket even as the amount of total outstanding paper plummeted.

This was what Jeff Immelt meant when he told Hank Paulson that GE was having a hard time selling paper for any duration longer than overnight.

The afternoon of Wednesday, October 1, 2008, Immelt talked on the phone with Tim Geithner, according to phone records published by the *New York Times*. Also on the hourlong talk was GE's general counsel Brackett B. Denniston III, among others.

Like most companies, GE had access to some cash through bank credit lines, an arrangement that allowed it to draw up to a certain amount at a determined rate. For many companies, bank credit lines serve as insurance against a cash crunch.

But in the financial crisis, when nervous investors feared any sign

of the next domino to fall, these bank lines weren't always a lifeline. When Countrywide Financial tapped its credit lines for $11.5 billion in the summer of 2007, it was a clear sign of desperation. With the questions swirling around the market, GE's bank lines were now effectively useless. Drawing on them would be the equivalent of shooting a flare into the sky and would almost certainly send investors, customers, and counterparties running away for fear that GE was in trouble. That could start a death spiral for the company. Even a healthy company can be destroyed by a runaway panic.

All this being said, in the minds of many, GE wasn't Lehman Brothers. It wasn't going to fail or have to declare bankruptcy. The extent of its military business alone made it unlikely that the government would allow GE to implode. The core businesses were considered strong — healthcare, jet engines, power turbines, and media — and they were all seen as solid bets for the future of America and the world. GE had real businesses owning real assets with real customers, and it was producing cash.

But GE was in fact in trouble. And it needed some help.

19

SUSPENDERS ON SUSPENDERS

DAVID MAGEE WAS starting to get the sense that he'd been stood up. Magee was a gregarious Mississippian business writer who specialized in books about corporate overhauls, turnarounds, and reinventions. He had written about thoughtful CEOs reshaping companies like Toyota, Nissan, John Deere, and Ford and was just the right guy, the image-shapers at GE thought, to write a story about Jeff Immelt that would remove him, at last, from the celebrity shadow of Welch.

Their timing could not have been worse.

GE handlers had given Magee access to executives like vice chairman John Rice and research head Mark Little, as well as time with Immelt himself. There was only one interview left on the writer's list, to be conducted in the sun-splashed office adjoining Immelt's at the company headquarters in Fairfield. Magee was to meet with Keith Sherin, the chief financial officer.

But on the appointed morning in the fall of 2008, Sherin canceled. Despite follow-up requests, the interview would never happen. As the pressure on GE mounted, Keith Sherin had somewhere else he needed to be.

The notion that had panicked investors and driven GE's stock price far below its true worth was a perfect pitch to one investor in particular: Warren Buffett, American business's patron saint of finding value where others had missed it. An investment from someone as well regarded as

Buffett might provide GE with a level of legitimacy that so far its own marketing machine had failed to secure.

In the chaos of the financial crisis — marked by extreme market swings, political uncertainty, and a turbo-charged rumor mill speculating on the next domino to fall — the financial services side of GE was simply too big to ignore. The solid fundamentals of the company — its massive industrial backlog of orders, its ability to generate cash, and its collection of hard assets — meant nothing to investors who had no desire to catch a falling knife.

In short, words didn't matter anymore. GE needed to show that the distressed markets were contorting its true value and that it had been wrongly caught in the torrent that was drowning others.

Warren Buffett, the billionaire investor with a folksy midwestern flair, was just the person to lend legitimacy at a time when GE couldn't convince people of its stability. He was also a businessman who understood that he wasn't just lending money when he made an investment — he was endorsing the target's fundamental strength. Buffett said that he wouldn't put money into a business he didn't understand, which was one reason he passed on helping American International Group Inc. (AIG) during the crisis before it was taken over by the government. His investment in GE would send a strong signal to the market.

But his investment came at a cost. Buffett, a veteran insurance investor who got his start handicapping horses, understood risk and dictated steep terms to reduce his exposure.

Just a week earlier, he had invested $5 billion in Goldman Sachs, another firm he thought was a victim of a mass hysteria that was distorting the value of possibly the best business on Wall Street. If Goldman had to pay, so did GE.

Buffett bought $3 billion in GE preferred shares, collecting $300 million in annual dividends, and he also received the right to buy $3 billion in GE common stock for $22.25 a share for five years.

The deal took only a few days to come together. For Buffett, it was a straightforward decision that he famously made early in the morning in his bathrobe from his house in Omaha.

It came at the right time. GE shares were down sharply, more than 10 percent that day, after a Deutsche Bank analyst came out with a negative assessment of GE earnings forecasts, citing the risk in GE Capital

and the potential trouble within. When the news of Buffett's investment broke in the middle of the day, it came to the rescue and had the exact impact GE management wanted: the stock recovered much of its value.

Besides the $3 billion vote of confidence from Warren Buffett, GE said that it would also sell $12 billion in stock in a public offering. This announcement was made only six days after Immelt's latest assertion that GE wouldn't need to make such a move.

For a company that depended on the credibility of its assurances and the trust of investors, it was sending a lot of mixed messages on Wall Street. The stock offering was no small undertaking, and it only confirmed the impression that GE was seeking a frighteningly large amount of cash on short notice.

The reversal was also expensive. GE had been repurchasing its shares right into the early days of the financial crisis, even as its shares were dropping. So far in 2008, the company had spent more than $3 billion buying its own stock. And in 2007, GE had spent $15 billion on its shares. Over the entire period, GE paid an average price of about $37.50 for half a million shares worth more than $18 billion. Now, it would sell almost 550,000 shares back to the market for $22.25 a share in order to raise $12.2 billion. By selling shares back to the market at a much lower price, GE was wiping out more than twice the amount of cash that the deal with Buffett had yielded. It was a disastrous use of the equity markets, and it wouldn't be the last time.

On October 7, the government brought GE some relief from the commercial paper crunch when the Federal Reserve created the Commercial Paper Funding Facility, which let companies buy three-month commercial paper from high-rated issuers. Paying a $100 million fee, GE signed up to join the program, which was scheduled to begin in less than three weeks.

Days later, GE reported its earnings and executives held a conference call that highlighted the steps taken to protect GE from volatility and economic disruption, including reductions in risk. Immelt insisted that GE's leaders had made moves that would enable the company to "outperform in any environment."

When it came to funding, GE never had a problem.

"Even with all this volatility we have never had issues in the CP market rolling our paper," Immelt told investors, assuring them that the

company was just being safe. The message was that the problems in the financial markets were very serious, but not for GE. As always, GE was an exception.

"I think we thought it was smart to have suspenders on suspenders on suspenders in this cycle," Immelt said, explaining the company's moves to batten down the hatches. And yet, within days, Immelt would be imploring the government to give GE access to insured funding.

A week later, in the middle of October, the Federal Deposit Insurance Corporation (FDIC) created the Temporary Liquidity Guarantee Program (TLGP), which would guarantee the debt issued by FDIC-insured banks, a qualifier that excluded GE.

Paulson called Immelt on a Sunday morning, according to the former treasury secretary, to explain the program to him and warn him that GE wouldn't get help. Paulson asked if it would be better for GE if the program happened or if it did not — and he was impressed with Immelt's answer.

"Maybe a lot of my guys would disagree with me, but the system is so vulnerable you should do whatever you can do, and we'll be better off than if it hadn't been done," Immelt said, Paulson later wrote in his book.

Immelt seems to have misspoken. Inside GE, the new program was seen as a real problem that could make their difficulties in the commercial paper market much worse. GE issued more commercial paper in the United States than anyone and was highly trusted by the counterparties buying the paper. But the stress and uncertainty in the commercial paper market during the crisis caused everyone to reduce their use of it — including from GE — and now investors could buy paper guaranteed by the US government.

What made GE attractive in times of financial turmoil, besides its reputation, was its triple-A credit rating. But this time was different. With storied Wall Street names going out of business overnight, there was simply no appetite among investors for risk backed by faith in any corporation. The FDIC program offset that fear as it tried to jump-start the market, but it also effectively blocked GE from the market because no investor would voluntarily buy GE's nonguaranteed paper.

Just a day later, Immelt was back on the phone with Paulson, telling him that his people were anxious about what this program would mean

for GE. "I'm worried about my company and our ability to roll over paper in the face of this," he said, according to Paulson.

Immelt was in Paulson's office two days later to argue that GE needed to be covered by the umbrella of the FDIC's program. GE was actually making the middle-market loans that banks typically didn't want, he argued, and now it was being punished in a dangerous way.

Paulson agreed, but needed to convince the head of the FDIC, Sheila Bair, a fifty-four-year-old Kansas Republican who spoke her mind to authority. She believed that taking risks brought rewards but also punishment. In other words, if someone made a bad bet, they should lose their money.

It was a pure interpretation of how markets should work and a major topic of debate around the financial crisis. The markets needed to have winners and losers. Intervening in the markets to help those who made bad decisions, no matter how calculated, distorted the risk taken by participants. That could have an impact on future investment behavior if risk was seen as having finite limits while returns had no corresponding ceiling.

Bair had fought to address aggressive practices in the subprime mortgage business and had waved warning flags at a time when few others were paying attention. During the financial crisis, which landed right in the middle of her five-year appointment, she asserted the power of her agency, which typically went about its work in a far corner of the financial regulatory framework. She had also shown a willingness to butt heads with Paulson and Geithner. Immelt went to visit Bair in person to give her the pitch for getting GE into the FDIC guarantee program.

In her memoir, Bair noted that the plan created a "huge competitive disadvantage" for GE because it allowed banks to sell FDIC-backed bonds. She mentioned Immelt's call and his visit to her office, noting that his request was supported by both his regulator, the Office of Thrift Supervision (OTS), and Paulson.

Bair asked for an examination from her staff. They found that GE's capital position, risk management, and information controls were favorable.

"I decided to approve it once Immelt agreed to have the commercial side of GE — the one that makes everything from lightbulbs to jet engines — guarantee us against loss," she wrote.

The FDIC changed its position to allow affiliates of FDIC-insured banks into the new program. GE just happened to own an industrial bank in Salt Lake City called GE Money Bank. It was FDIC-insured. GE was in.

After repeatedly asserting that it didn't need help, GE would use the government guarantee to sell almost $131 billion in debt through a staggering 4,328 different issuances. The second most active issuer, Citigroup, had just 1,655 issuances.

20

CIRCLE THE WAGONS

I F GE'S STOCK price was a measure of investor perception, it was heading in the wrong direction. They were deep into the worst valuation of Jeff Immelt's time at the top. On Immelt's first day in 2001, the shares were around $38. When he was dining in Beverly Hills in September 2008, the stock was at $25. Shares opened 2009 at around $15 and closed the month of January at $12.

Always aware of the stock's reflection on his leadership, Immelt was trapped in a waking nightmare, and the constant pressure of the crisis was taking a toll on him. He was also increasingly frustrated by the semipublic disdain emanating from Jack Welch, who let every intimate acquaintance know how badly he thought his old shop was weathering the storm.

Immelt finally let his displeasure fly at a Manhattan event, the account of which flew directly into a column in the *Financial Times*. Jack Welch had benefited from the red-hot economy of his era and had never been tested by a major macroeconomic crisis, Immelt told the crowd. Immelt's claim that Welch had run GE with economic winds at his back was fairly uncontroversial. But Immelt continued.

In the 1990s, "anyone could have run GE and done well," he quipped. "Not only could anyone have run GE in the 1990s, his dog could have run GE, a German Shepherd could have run GE."

Immelt's representatives denied the jab, to little effect, as it raced from the *FT* around Wall Street and Midtown Manhattan. It was a startling

sign of bitterness from the usually genial Immelt, especially about a figure still revered within the company and its investor base.

The dawn of 2009 brought hope for a better year. GE ended the previous year with an investor presentation on GE Capital. Executives took deep dives into the businesses and answered questions. GE also announced that it would no longer give quarterly earnings projections, partly because it recently had a hard time making those goals.

The meeting seemed to restore some calm among investors, but it didn't last. The stock continued its downward trajectory as the financial crisis roared on. By the middle of February, GE shares were trading in the single digits.

The financial crisis was putting a full squeeze on GE, and Immelt conceded in his annual shareholder letter that some of its holdings were a problem. He flagged GE's plunge into commercial real estate, a portfolio that had ballooned as recently as 2007.

"Did we end up with too much exposure in certain areas during the credit bubble? Maybe, a few," he wrote. "Today, I wish we had less exposure to commercial real estate and UK mortgages."

Chatter about GE's financial condition continued. Investors were worried that a cut to the dividend was coming and that the rating agencies were going to downgrade GE's sterling credit rating. Many other large companies were reducing dividends, including Dow Chemical, Pfizer, and Citigroup, but shareholders were worried that GE's stock would drop sharply if that happened. The solution to the uncertainty was to sell the stock, but that would just feed the cycle.

Immelt, ever the believer, stood strong and scoffed at the idea that GE would need to cut its dividend. "We've got the cash flow to pay the dividend," he said in a February 5 interview with the *Wall Street Journal*.

The concern over the dividend had become a heavy weight on the company. Given its distressed state and reliance on government help, GE needed to preserve cash, and the first step was cutting the dividend. But that was absolutely unthinkable to Immelt.

It had been more than seventy years since GE last cut its dividend. For Immelt, maintaining that track record was a sacred duty. He saw it as so important to shareholders that reducing the regular payout of cash to loyal investors might permanently alter GE's valuation. He also knew that cutting the dividend would be a permanent stain on his own record.

It wasn't a comfort that other companies made the same prudent move in times of crisis. GE was supposed to be the exception.

The moment came on February 27. Immelt cut the quarterly dividend to 10 cents from 31 cents.

Immelt would for years after refer to that day as his worst ever on the job. The cut would not only hit investors but also cut the quarterly checks going out to employee shareholders and retirees.

Wall Street cheered the move. When a company is stretching to pull cash from anywhere it can, cutting the dividend is the prudent move. Generous dividend payouts — like share repurchases — don't make sense for distressed companies. By cutting the dividend, GE removed uncertainty and also saved billions of dollars in cash going out the window every quarter.

For thousands of longtime GE investors, whose modest shares had regularly produced a payout of cash every three months, the company's gesture to the street was far more distressing. A significant if not primary source of income had been abruptly slashed, with no warning or recourse.

After the cut, the early indicators were promising. GE's stock kept falling for a while, but then it stabilized and turned around. It would close, at its lowest point in the crisis, at $6.66, but then rebound above $10 within weeks.

In a further attempt to stem the tide, GE decided to hold another investor meeting on GE Capital. The deep dive at the meeting would take the form of an exhaustive internal review. Capital executives spent five hours explaining all the parts of the division and how it made money, trying to win back the favor of a skeptical crowd of Wall Street analysts.

It is easy to forget why it was so important for GE to stem the tide where it could, to forget what was at stake for the company in that moment. If the decline of GE's stock price didn't stop, there was no telling how the drama might end, as the disappearance of other iconic companies in recent months had made brutally clear.

New York Times business columnist Joe Nocera wrote about GE's situation near the stock price lows of the crisis in March. Nocera has shown himself to be a fan of Immelt. Less than two years earlier, he had written a rave tribute to Immelt's greatness in pioneering a new type of leadership and approach to business. But Nocera wore a very different

hat in March 2009, when he blasted the company for its lack of credibility and uncertain financial condition. He detailed Keith Sherin's recent insistence that GE didn't need cash and slammed the company for demanding that investors trust its numbers and "the rock solid image of GE" on faith alone.

"You couldn't help thinking that we'd seen this movie before," he wrote, comparing GE to other companies that came under intense pressure in the crisis, like Bear Stearns, Lehman Brothers, AIG, Merrill Lynch, and Citigroup.

GE Capital didn't mark most of its assets to market prices, openly distorting GE's financial position at a time when the market was severely weakened. Nocera said that the only way out was for GE to detail everything in the GE Capital investor meeting to show investors what they should or shouldn't be worried about.

"But no investor is going to take Mr. Sherin's word for it — not anymore," Nocera wrote.

GE's triple-A credit rating remained a concern as well. Having the stellar rating had been the key to GE's success for decades. The company could borrow cheaply to fund manufacturing and raise money for GE Capital to lend.

On March 12, S&P cut its credit rating on GE, and Moody's did the same a dozen days later. The downgrade was a painful dose of reality, but not a surprise. Given its portfolio in financial services, the credit rating was too high. Rating agencies will often give firms the benefit of the doubt for a period before downgrading, but GE couldn't hold on any longer.

The agencies cited the obvious concerns that weren't in line with expectations for a top-rated company. GE Capital had $600 billion in assets with massive risk, including a $28 billion private-label credit card operation, $22 billion in residential mortgages, and $81 billion in commercial real estate.

GE downplayed the rating downgrade, saying that it confirmed the soundness of the company. There was no question, however, that the downgrade was a historic blow to the psyche of the company and its mythology of excellence.

The GE Capital meeting finally arrived on March 19. The daylong affair in a chilly hotel conference room in New York featured 176 Power-

Point slides and hours of dry explanation of the risks and safeguards of the Capital businesses, presented from a long table of executives led by Neal and Sherin.

Two hours into the intense meeting, there was an interruption when the fire alarm went off and the building was evacuated. When everyone returned to the conference room, the GE executives resumed their endless drilling into Capital's massive balance sheet and all the risks that the credit rating agencies were fretting about. They were making the case that GE had properly marked its assets and was managing the situation prudently.

It worked. After all its efforts to calm investors, the one tactic that worked was raw transparency. Investors and analysts generally felt better, assured that GE had finally revealed its secrets. The presentations hadn't been completely exhaustive — there are a lot of places for trouble to hide amid more than $600 billion in assets — and there was still plenty of risk in the portfolio. But clearly explaining to investors the deeper details of the business helped to make them more comfortable with the adequacy of GE's balance sheet.

For GE, it was the moment that finally started to put the financial crisis, ever so slightly, in the rearview mirror.

To Beth Comstock, the shift in how commentators talked about GE highlighted the need for a "sense-making project": gathering information that would help GE figure out what had turned people away from the brand, and what might reverse those negative associations. The GE marketing machine acted by soliciting both sides of the just-concluded presidential contest, in which Senator Barack Obama, the Democratic candidate, had walloped Republican Senator John McCain. Comstock brought in Steve Schmidt, who had run McCain's campaign, and David Plouffe, Obama's campaign manager, to advise the company on how to address the mistrust of the public.

Schmidt's advice would come to define the methods of Comstock's media and public relations operation. The campaigns of both politicians and companies, Schmidt told GE, were not "won by the candidate or company with the best character, or product, but by the one with the simplest and most clearly told story."

In Comstock's summary: "Pick a simple story . . . and tell it again, and again, and again."

21

KABLETOWN

G E STILL NEEDED cash in 2009, and it would turn to one of its most prized assets to get it.

In 2009, after owning NBC Universal for twenty-three years, the company began to shop around the once-high-flying business that Welch and Immelt wore like a sparkling gem. Owning NBCU gave them access to red carpets and fancy parties, and was a helpful tool for GE's marketing efforts.

In addition, GE's ownership of CNBC — a ubiquitous presence on Wall Street trading desks — gave it a direct pathway to its investors as morning anchors fawned over GE management's on-air appearances. The cable networks were doing well, but unfortunately the broadcast network wasn't; its declining viewership was exacerbated by an advertising downturn.

Ever since he took over, Immelt had been trailed by questions about the potential sale of the television unit, which some saw as a vanity acquisition that had really never belonged in an industrial company. Without fail, he shot down the idea and always defended the quality of the business.

But as GE emerged from the financial crisis, it needed to do a deal to get cash, and some saw selling NBC Universal as the prudent move, even if the timing wasn't great because of industry valuations. Selling the unit now made the most sense to Immelt as well. If nothing else, the sale

of the television division would bolster his argument that he was making GE simpler and more manageable.

Others saw the sale, however, as more evidence of Immelt's rudderless strategy. He had piled on assets by acquiring Universal from Vivendi and, perplexingly, teamed up in mid-2008 with private equity firms to shell out a staggering $3.5 billion for the Weather Channel. But now GE was moving to dump substantially all of these media assets.

In truth, Immelt had shown some caution in chasing deals. In the summer of 2007, the pursuit by media mogul Rupert Murdoch of an acquisition of Dow Jones & Company, owner of the *Wall Street Journal*, created a rift in the Bancroft family, which had controlled Dow Jones for more than a century. Murdoch had offered $5 billion to buy Dow Jones, a whopping 65 percent premium, but some in the Bancroft family disliked the Australian billionaire and worried that he would turn the *Journal* into a tabloid. A search for alternatives began.

Immelt was skeptical at first, but there was an argument for merging the storied financial news company into CNBC to create a financial media powerhouse.

"You'll be the biggest swinging dick on Wall Street," barked Jack Welch to Immelt. The former CEO, who had always paid close attention to CNBC, his pet project, saw an opportunity to own the two biggest names in finance.

Immelt gave it some more consideration, and GE's deals team got to work. Eventually it joined together with Pearson PLC, owner of the *Financial Times,* to discuss making a bid for Dow Jones. But an offer from Immelt never came.

The deal simply didn't make sense for GE. Just getting in the door required topping Murdoch's $5 billion, and Dow Jones & Company wasn't worth anywhere near that amount, GE's deals team determined. Also, Murdoch made it clear that he had long aspired to own the *Journal*. Immelt had no interest in getting into a public bidding war with Rupert Murdoch in his pursuit of the jewel in his crown.

In addition, as it came out of the financial crisis, GE wasn't interested in spending more on media deals, even though the NBCU leaders thought that shifts in the media landscape made such deals necessary. Broadcast was fading, cable was rising, and video streaming was com-

ing. Immelt had little interest in the wheeling and dealing necessary to manage that change.

So GE shopped NBCU around, holding talks with Time Warner and even Murdoch's News Corp. After nine months, Philadelphia cable giant Comcast won the day. GE agreed to sell to Comcast a 51 percent stake in the division at a valuation of $30 billion. The remaining stake would be sold over a number of years.

That sale price was shocking to one person: Bob Wright, the former vice chairman and CEO of NBC Universal. Just three years earlier, a meeting with David Rubenstein, founder of the private equity firm Carlyle Group, had yielded a potential deal for NBCU. The two came up with the outlines of a buyout that would separate NBCU from GE, with Wright still running it. The valuation of the proposed deal was $45 billion.

That deal had been blocked immediately by Immelt, who had insisted that the business wasn't for sale. Wright clearly saw that GE was weathering the economic downturn by focusing on its traditional industrial business and not investing in NBCU. He decided to retire within months of the decision. As Wright saw it, by the time Immelt reversed course and finally cut a deal with Comcast, much of the value that GE had worked to create in NBCU was lost.

A year after rejecting Wright's proposal, Immelt had gone on Charlie Rose's television show and again faced the question of whether GE might sell NBCU. Immelt had dismissed the idea immediately while also revealing that he apparently had never given much credence to Wright's buyout idea or sought the counsel of his overseers.

"I've never had one serious discussion about selling it. You know, I've never talked to the board about selling it," he told Rose.

When it did finally happen — news of the talks broke in the fall of 2009 — the sale was generally cheered by Wall Street, as the media assets had always been an odd fit with the rest of the company. There were the usual questions about the deal timing, regulatory scrutiny, and price. GE seemed to be selling NBCU into a poor market where the internet was beginning to upend the traditional model. The NBC Universal division had no profit growth from 2005 to 2008 and was down almost $1 billion by the middle of 2009.

In the wake of the financial crisis, GE's sale of NBCU was reassur-

ing to investors when placed alongside the effort to move away from financial services. When GE had bought RCA in 1985, giving it control of NBC, the conglomerate's largest units had been power, aircraft engines, and technical products, including medical equipment. Now, in many ways, the company seemed to be moving back to its roots.

22

NO OTHER WAY

JEFF IMMELT TRAVELED up to West Point in December 2009 to give a speech entitled "Renewing American Leadership."

The historic military academy is about fifty miles up the Hudson River from Manhattan and not too far from GE's Crotonville compound. Appropriately, it was the weekend of the Army-Navy football game, and Immelt opened his speech by making a crack about his college days as a lineman and lamenting that since he was no longer eligible to play he wouldn't be able to help Army win the game that day.

It was prime Jeff Immelt. He then ran through his usual dramatic talking points, hitting on GE's importance in the world and the struggles it (he) had faced, including 9/11, the Iraq and Afghanistan wars, and economic turmoil. With an eye toward economic stimulus, he also spoke of partnering with the government on massive energy and manufacturing spending.

Facing the military's leaders of the future, he opened up about the recent financial crisis and the road ahead in signature Immelt fashion — by apologizing for his and GE's failings:

"I decided that I needed to be a better listener coming out of the crisis. I felt like I should have done more to anticipate the radical changes that occurred."

And then he was quick to chastise others for their greed and pointed a stiff finger at the financial services industry: "While some of America's competitors were throttling up on manufacturing and R&D, we de-

emphasized technology," he said. "Our economy tilted instead toward the quicker profits of financial services."

He spoke in a way that Immelt had mastered. He acknowledged some failings, while also distinguishing GE from its lesser competitors. In this case, he brought out a familiar argument. GE Capital, one of the biggest financial services operations in the industry that he was denouncing, was in fact nothing like a bank, and its operations were balanced by an unsinkable industrial powerhouse. It was other banks that had let financial services take over the economy, and their actions that had harmed others. Again, GE, in its telling, was the exception.

"Rewards became perverted. The richest people made the most mistakes with the least accountability. In too many situations, leaders divided us — instead of bringing us together," Immelt told the West Point audience. He then returned to the theme of what set GE apart from its peers by giving the cadets his sage thoughts on how "it takes courage to rethink your leadership paradigm."

As he often did, Immelt spoke with a combination of boasting and joking. He was a pro at rattling off a very serious story while cackling to give it a lighter feel. Speaking of his leadership during the financial crisis, he gave his audience an unintentionally candid glimpse into his unilateral decision-making during one of the most chaotic times in recent financial history.

"I had weekly board calls, making frequent decisions, and doing things I never thought I would have to do. I am sure that my board and investors frequently wondered what in the heck I was doing."

On the surface, Immelt was taking credit for GE's navigation of the crisis and highlighting his bold leadership. But he also admitted that the board of directors was only peripherally involved in his decision-making. Immelt, who was both CEO and chairman of the board, wasn't seeking constant advice or approval; instead, he was moving quickly, backed by the board's trust in him. On the other hand, he was ignoring the prime reason that companies have an independent board: to protect the investors who elect them and help manage risk.

"I had to act without perfect knowledge; I had to act faster than my ability to communicate or explain my actions. I could do this because we had built trust. And we kept GE safe because we moved fast."

Of course, GE was also kept safe because of the helping hand of the

federal government, and that came at a price. As Immelt lectured cadets about integrity and the future of leadership, his own future inevitably was shaping up to include more and more regulators looking over his shoulder.

But first, GE had to deal with its past. After more than two and a half years, the Securities and Exchange Commission concluded its investigation into GE accounting practices, having found multiple instances of misbehavior in the pursuit of financial targets.

The company had overstated its earnings by hundreds of millions of dollars and stretched the accounting rules to their breaking point. For its sins, GE agreed to pay a $50 million fine and essentially promised not to misbehave again. The fine was small considering that the SEC had alleged violations that were intended to present the company's results as something they weren't. In the end, GE managed to keep up the appearance of stability and ensure there were no surprises for investors.

The settlement seemed to confirm the long-held suspicions of accounting games within the company. Many on Wall Street scratched their heads at the relatively small fine, which seemed unlikely to deter future misbehavior. The SEC's charges included multiple counts of fraud, filing misleading documents, and failure to maintain accurate records and internal controls.

The SEC settlement came as sweeping reform to the financial services sector was taking hold of an industry already saddled with heavy, if ineffective, regulation. After all, those who almost destroy the entire financial system have to expect some changes as to how they are allowed to conduct business.

The sweeping changes proposed that summer by President Obama formed the basis for what would become the Dodd-Frank Wall Street Reform and Consumer Protection Act. Signed into law in July 2010 by Obama, the act effectively overhauled the entire system with a promise that companies would never be bailed out again using taxpayer money.

Dodd-Frank marked a tectonic shift in how banks were allowed to do business. The two-thousand-page law rolled back risky strategies at large financial firms and changed everything from credit cards to mortgages. The balance sheets of the major banks would get regular check-

ups from the government. Yearly stress tests would determine whether banks had enough cash to cover extreme distress.

Of course, GE was different. It wasn't technically a bank. For years, GE Capital was regulated by the Office of Thrift Supervision, a division of the Treasury Department established in the 1980s in the wake of the savings and loan crisis. The OTS wasn't expected to survive the coming reform, and it didn't. The agency had overseen Washington Mutual, IndyMac, and the financial products division of AIG, all of which collapsed or came very close to collapsing. The mortgage giant Countrywide Financial had reorganized itself in 2005 in a way that shifted its regulator to the OTS, a move that some saw as a scheme to get a less stringent regulator. Countrywide collapsed during the housing crisis and was acquired by Bank of America in a fire sale.

With the OTS dissolved, GE's regulator would now be the Federal Reserve, which would regulate it like a bank. An army of Fed officials would now set up shop in GE Capital's offices in Norwalk, stress-testing the company's financials.

After the near-death experience of seeking federal help, and with new regulators on GE's back, Immelt was openly talking about shrinking Capital when the time was right. Years would pass before GE made a major move out of Capital, he said, but "at some point capital markets are going to open up and we are going to be able to do some bulk dispositions at that time."

In other words, when the door opened, he planned to push right through it. Immelt seemed to have finally had enough of the division that plagued him from the start. The crisis had made the internal separation between the finance team and the industrial businesses even wider.

GE Capital employees at management training sessions in Crotonville flipped their name placards around in some meetings to hide their division. It helped them avoid the scorn of those from other businesses who blamed the finance cowboys for almost destroying GE and punishing the legitimate industrial operations.

In the meantime, even though GE seemed to be planning to shrink Capital after learning a humbling lesson, the deal-making didn't stop. In mid-2011, it made an unexpected cash offer for ING's US online banking

business. The thinking was that the bank deposits would be an attractive form of funding for its other lending and leasing operations, and a cheap alternative to commercial paper. But GE's offer didn't win the day, and the business was acquired by Capital One for $9 billion.

Capital begrudgingly accepted the presence of federal regulators, but the tension in its offices was clear. GE didn't like outsiders telling it how to better run its operation — that was typically GE's role. In May 2012, GE Capital's leader, Mike Neal, made it clear that the new regulatory atmosphere was a big change, but said that people were learning and the relationship was "constructive." He estimated that the Fed's presence was costing GE about $400 million a year, but he seemed to have made peace with the situation.

"They are going to be with us for a very long time," he said. "There is really no other way to think about it in that regard."

23

GREEN IS GREEN

WASHINGTON, DC, WAS unseasonably warm in late September 2010, when Jeff Immelt and a pair of company attorneys arrived at the William Jefferson Clinton Federal Building, the massive neoclassical home of the US Environmental Protection Agency.

GE and the EPA were locked in a long and bitter feud that stretched back decades over who was responsible for cleaning up the toxic mess in New York's Hudson River. A pair of GE factories had pumped in more than a million pounds of toxic compounds for generations, turning one of the most famous and vital American waterways into a two-hundred-mile-long Superfund site.

The mere mention of the Hudson to Welch had been enough to set off a finger-pointing screed. Polychlorinated biphenyls, or PCBs, had been widely used in industry for decades. The compounds were found in lubricants and in the heavy-duty, fire-resistant insulating fluid inside the electrical capacitors that GE churned out of factories like those in Hudson Falls and Fort Edward, New York.

And it had been legal at the time, Welch would shout, to pump the excess fluid into the river. He was right that it was legal at first, but eventually, as study after study about the harmful effects of the chemical piled up, it stopped being legal. PCBs would eventually be classified as a likely human carcinogen. GE stopped dumping them in 1977, two years before PCBs were banned altogether. But PCBs still permeated the grounds of

those factories and trickled from the sides of the Hudson Falls plant into the river for more than two decades.

After battling for thirty years, GE and the EPA were arriving at last at a moment of decision.

The GE contingent ascended to the building's third floor, where they entered the Bullet Room, a long, wood-paneled conference room used for important meetings. The group included GE environmental lawyer Ann Klee, who had been hired largely because of her past service at the EPA, and the company's patrician general counsel, Brackett Denniston III. Across the table sat a cadre of EPA officials led by Lisa Jackson, President Barack Obama's first EPA administrator and a veteran environmental engineer and regulator.

To some in the room, the GE team seemed to think that they had sewn up a Hudson River deal to their liking. The company proposed capping the largest "hot spots" of contamination by sealing them with a bed of clay and soil. The GE team was in for a rude awakening.

The EPA and some local government officials hated this idea. Over time, they said, the river would slowly eat away at the caps and unleash the chemical. Jackson told the GE team, cordially but firmly, that GE would be responsible for digging up and removing the toxins from hot spots along some forty miles of the riverbed, the very remedy GE had hoped to avoid. The cost of disposing of the sludge also fell to the company, bringing the expected bill to around $2 billion.

Ann Klee, the GE lawyer and former EPA official, exploded.

"This is the wrong decision! We're going to meet you in court," Klee exclaimed. While the EPA might try to dictate how the second phase of dredging should play out, Klee said, "We're not going to follow it."

"She kind of lost it," a person who attended said. "They thought they had a deal that didn't exist."

Denniston, his chummy glow gone, was silent. Immelt cut in, silencing Klee, playing the good cop. "Okay, we don't have to decide this today," he said.

Jackson coolly corrected him.

"Oh no, we're deciding today," the EPA administrator told the CEO. "This is done."

The meeting ended quickly. As the GE executives departed, the only question that remained was whether the company would go along with

the EPA's plans — and foot the bill — or plunge the question of the Hudson River cleanup back into limbo by fighting the EPA in court.

It had been almost a decade since Immelt succeeded Jack Welch, and General Electric had come a long way in the public image it presented about environmental matters.

In 2002, the recently retired former CEO, clad in a cable-knit sweater, had raved in a televised interview with Bill Moyers that the EPA's initial plan to dredge the bottom of the Hudson was "just nuts" and amounted to an attempt to "chase molecules in water."

Rocking and gesticulating, Welch exclaimed, "The Hudson's recovered, it's wonderful, it's beautiful," before adding, in an aside that seemed to take Moyers aback, that he didn't even believe that PCBs had been proven harmful to human health in the first place. The producers cut to experts explaining that multiple studies had found the compounds to be probable human carcinogens.

Jeff Immelt didn't feel like paying billions to clean up a river either. Yet he was smart enough to see an advantage in updating GE's public image on the environment. The company, for all the prowess of its engineering, tended to come off as an implacable ecological bad guy, and that didn't fit with the sunny salesman's idea of the GE image. He might at least avoid some of the confrontations that Welch had been drawn into as he fought against the environmentalists and lawmakers insisting that GE clean up the Hudson and other rivers that its factories had despoiled, especially the Housatonic in Massachusetts.

The lowlight came in late April 1998 in Cincinnati, the site of GE shareholders' annual gathering. During discussion of a shareholder proposal for GE to do more about the polluted rivers, Welch dismissed the idea and declared that there was no danger to the public. One shareholder challenged Welch on that assertion, arguing that GE's aggressive interpretations of scientific studies were comparable to those of tobacco companies trying to dodge responsibility for lung cancer. Welch exploded.

"That is an outrageous comparison," an exasperated Welch exclaimed from the dais to the activist. "You owe it to God to be on the side of truth here." Another shareholder yelled, "Sit down," producing a round of applause.

To the consternation of GE's public affairs office, the speaker being ridiculed was Sister Patricia Daly, a Catholic nun from New Jersey who had pushed for socially responsible behavior by corporations. The ruthless backlash became a symbol of GE's relentless opposition to the issue and painted the company as hostile to environmental issues.

Beyond the issue of the New York river, there was the larger question — to GE anyway — of its brand. A carefully curated self-mythology had always been central to GE's methods, going back to Edison. GE had marshaled the tools of public opinion that had marked every era — hence the cameos of Ronald Reagan and Kurt Vonnegut in its corporate archives and its sponsorship of general entertainment that stretched far afield from its direct interests, such as early live TV. When the company's public relations efforts were working right, the result was a general good feeling about the company. Immelt was fond of citing the conclusion of a study that the brand might have been worth $50 billion — a sign of both its worldwide recognition and the impression that, whatever the reason it popped up in the news or at a congressional inquiry, the company was a vast organization of elite engineers who made vital equipment that helped the world run well. But this legacy brand advantage was shifting and changing in the twenty-first century, and not just because of the rising and reasonable skepticism about whether major public companies were doing good things for the right reasons. The pursuit of profit for shareholders had left a lot of ethical and moral gray areas.

GE was also in a long, slow process of changing its relationship with individual American consumers that had begun under Welch himself. When GE's "We Bring Good Things to Life" was coined, the company could boast of selling Americans their radios, televisions, microwaves, stoves, refrigerators, air conditioners, and incandescent lightbulbs. Welch sold off units like these, such as the small appliance business, which made toasters and small kitchen goods. The few of those business lines that remained in GE's industrial portfolio by the mid-2000s were essentially marked for disposal once both the market and the time were right. In fact, under Immelt, the quest to shed those humdrum legacy products was only accelerating.

Partly, shedding these units was the company's natural response to an era of free trade and rising global competition in which American gad-

get manufacturers, especially those on the commoditized low end, simply couldn't compete. But Immelt wasn't exiting low-margin businesses like kitchen white goods and lighting to enter other sectors more prized by American consumers. The era of GE as a consumer-facing brand was ending. The meatball logo retained much of its power, still calling out from the aisles of hardware stores and appliance retailers, but thanks to quiet licensing deals made over the years, the logo could be found on gadgets that GE hadn't actually made for decades, or had never made. (Among these products were a GE-branded computer mouse and a faux security sign meant to fool burglars into believing that a house was protected by a fictional GE alarm system.) Now more than ever, the company's core customers would be institutional: hospitals, the military, utilities and power generation outfits, airlines.

As GE evolved into a more business-to-business company, it didn't lose the need for popular support. A positive public understanding of a company like GE — that it was familiar, competent, homely, and ultimately good — would remain essential in rebuffing the untold misfortunes and scandals that could arise at any moment for such a huge and complex enterprise. Hence, the company needed to find another source of warm and fuzzy feelings about GE besides the familiarity with the brand customers felt as they plonked down a few bucks for a two-pack of 60-watt lightbulbs.

GE had been demonized before, after all. The company had suffered through college campus boycotts in response to the huge sums it made contracting with the US Department of Defense. It had been battered in the press following the tsunami-driven failure of nuclear reactors at Fukushima in Japan, a crisis that blindsided the company, which had produced those reactors four decades earlier; only a few people who even knew how they had been designed were still employed at GE. And though GE had licensed its logo for forty years to Haier to use in selling stoves and washing machines, that brand presence for GE represented only a compensated risk: a wave of exploding ovens could hammer the stock price, even though GE no longer had anything to do with building kitchen goods.

It helped in such instances to have seeded the mind of the public with a general idea of the company's good intentions and competence — to have been nurturing a warm brand glow. And what GE needed to do

this, in the eyes of Immelt and some of his closest deputies, was a story. The narrative of success begat success, and the imagination of excellence and innovation was the first step to achieving both. One of those Immelt allies would later put it succinctly: "Strategy is a story well told." That executive was GE's chief marketing officer, Beth Comstock.

General Electric's pressure campaign against the EPA over the Hudson River had not just been waged in Washington. It had been fought most fiercely on the ground in Hudson Falls and Fort Edward, New York, two former GE factory towns full of loyal employees and retirees, as well as residents devoted to exploring and fishing in the river. The company funded websites, took out ads in local newspapers, aired infomercials, and broadcast TV spots that featured ominous footage of what it depicted as the EPA's plan for the scenic waterway of the Upper Hudson: ruinous clamshell diggers and truck traffic, making a mess of the river bottom in the name of a cleanup. "You've just experienced twenty seconds of dredging," one said, over shots of huge earth-moving equipment. "Imagine what twenty years would be like."

The ads worked. Public resistance to dredging plans rose in the very communities most directly inundated by GE's pollution, albeit with significant opposition from some who felt that the company's efforts were heavy-handed and misleading.

"These people are really, really scared of what this project is going to do to their communities," an environmental activist told a New Jersey legislative committee in June 2001 even as lawmakers hundreds of miles downriver from GE's PCB pollution sites tried to pressure the company to go along with a cleanup. "Why are they afraid of what it's going to do? They are afraid because General Electric's public relations people have been telling them to be afraid. They're afraid because General Electric has been spending millions of dollars to avoid culpability."

The mainstream television ads that would come to define GE's public image in the Immelt years took an entirely different approach. Whereas Welch, and Immelt in his earliest years, had pressed the company's interests on the environment in the smashmouth tone of a congressional race, the tone of Immelt's tenure, now set by Comstock, was imbued with the upbeat narrative stylings she had imported from NBC. Those stylings meshed perfectly with the boosterish sunny quality of Immelt's

style in telling the public gently, cutely, and vaguely how to think good things about the world's most famous industrial company.

Immelt himself debuted a change of corporate tone when he unveiled a slate of new GE promises to hit environmental goals in a ballroom in Washington, DC, in the spring of 2005. On screens in the hall, a tranquil rainforest image appeared, along with the strains of a familiar tune as the action began: a CGI-animated elephant calf dancing through the jungle to "Singin' in the Rain."

"Water that's more pure," a narrator breathily intoned as the elephant tap-danced through a puddle. "Jet engines, trains, and power plants that run dramatically cleaner."

The ad announced a new coinage of Comstock's, produced out of one of the "imagination breakthroughs" she led with executives to launch new marketing campaigns and branding for the company. "Ecomagination," GE said, was an effort to "create technology that's right in step with nature."

Ecomagination, the result of months of work and deliberation, had met with a fair amount of internal resistance to any attempt to rebrand GE as an ecologically minded entity, Comstock later wrote in her memoir. "You're going to make us look like idiots!" one executive had bellowed upon a first viewing of the "Ellie the Elephant" ad.

The truth was that GE was hedging its bets and casting its lot with other businesses that had decided it was good for business to be seen as interested in protecting the environment. But "ecomagination," in addition to being a mouthful, was also a bit of a stretch. GE did commit to concrete goals. It would cut its own greenhouse gas emissions and raise investment in research on "cleaner technologies" from $700 million annually to $1.5 billion by 2010, the company said in a press release. The company also aimed to double — from $10 billion to $20 billion by 2010 — the revenue it generated from products and services that "provide significant and measurable environmental performance advantages to customers." An independent environmental organization had been consulted to determine which GE offerings could qualify for inclusion in "ecomagination."

But the initial list of seventeen products that Immelt boasted of that spring day in Washington showed that change was in the eye of the beholder. The offerings GE singled out as proof of its environmental con-

version included: a new jet engine, called the GEnx, that had already been years in development and whose fuel efficiency would improve on the fuel efficiency of the older engines it replaced; the GE Profile Harmony washing machine, whose name had come from being "in harmony with the environment"; and Evolution, a new series of GE locomotives for the freight railroad industry that would feature greater fuel efficiency and cleaner emissions than the smoke-belching machines that came before. "Go ahead," the promo copy for the locomotives encouraged, "take a closer look at this product of pure ecomagination."

Was it, though? GE had hardly changed its business model in designing the GEnx engine to use less fuel. The engine was designed at the behest of Boeing, which wanted new power plants for its latest model 747 and new 787 Dreamliner. Those planes would be marketed to airlines that measured fuel by the drop as they struggled to eke out profit margins that were directly affected by the efficiency of the planes they flew. It was good news for the greater goal of reducing global carbon emissions that cost-consciousness was driving the aviation industry to seek greater efficiency (though greater efficiency was counterbalanced in part by a top priority of engine-makers like GE: to keep the number of global flight hours steadily rising so as to turn their engine fleets profitable). But GE's efforts to improve jet engine fuel efficiency were hardly a result of a green epiphany among General Electric's top executives.

As for the Evolution series of locomotives, the first preproduction units had gone out on the rails in 2004, a year before Immelt debuted ecomagination and the elephant danced. With substantial improvements in their emissions, the locomotives would be an enormous commercial success: hundreds of units were sold to freight railroads and mining companies around the world. But were they a "product of pure ecomagination"? The Evolution locomotive was, if anything, a product of the EPA, whose tightened emissions standards had compelled railroads to find locomotives less damaging to the environment — and thereby created for GE the incentive to invest in designing and producing them.

Nonetheless, the "imagination breakthrough" session had yielded a slogan that yielded GE a flood of mostly positive press clippings. Now the push was on. Ellie the elephant hit the airwaves, as would a series of

cheery commercials in the months to come, emanating from GE's long-time ad firm BBDO Worldwide, which pressed the corporate image as a font of industrial innovation, animated by thoughts of saving wild animals and the planet.

In one spot, workers assembling a power turbine break into a line dance; they are joined one by one by others on the factory floor, and then an unbroken stream of employees and customers around the planet join the line. It was like a Coke commercial with jet fuel and ecological wish-casting at its heart. In another, which the company called "Model Miners," semidressed men and women strut about in headlamps in a coal mine, sensually brandishing jackhammers and picks, to the strains of "Sixteen Tons," the American folk classic about miners sinking into debt. The punch line was a reference to "clean coal": "Thanks to emissions-reducing technology from GE Energy, harnessing the power of coal is looking more beautiful every day."

Some commentators noted the incongruity of using a song about the miseries of the twentieth-century mining life ("I owe my soul to the company store") to extol the beauty of the coal business. There was also the usual tell of an appeal to the flesh: "One thing is clear, though," Joshua Ozersky wrote in the *New York Times*. "Throwing oiled-up hotties at the camera is, in commercials as in hip-hop videos, usually a sign you don't have anything to say."

Some environmentalists who had fought GE for decades, not just on the Hudson cleanup but on the company's broader track record on pollution and emissions, were appalled. But ecomagination mostly worked. The clips unspooled in a wave: old GE was turning over a new environmental leaf and pledging to make money doing so. *Forbes,* for instance, weighed in that August ("GE Turns Green"), albeit with a nod to the fact that some of ecomagination amounted to "showmanship." (*Forbes* was skeptical of GE's apparent capitulation to environmentalists, but it did like the models.)

"In its essence it's a way to sell more products and services," Immelt told the magazine. On another occasion, he said simply, "Green is green."

And with the rising of one of GE's most ambitious marketing campaigns, the star of Beth Comstock rose as well. She had been named GE's first chief marketing officer in a decade. She was headed higher.

• • •

On December 16, 2010, Lisa Jackson had a ten-minute hold on her calendar for a call with Jeff Immelt. GE had pecked away at the EPA and its dredging proposal, seeking alterations in the government's plan that would limit the amount of dredging GE would have to pay for in the Upper Hudson — anything to hold down the cost of the cleanup that now seemed inevitable. Inside the EPA, officials waited anxiously to hear whether GE planned to fight the proposed cleanup plan in court. A lawsuit might only postpone the inevitable for the company, but every year of postponement was another year of accumulation, the PCBs building up in the tissues of the wildlife in the Hudson. Still, the agency held firm, refusing to make changes to its proposed remedy. Now Jackson was calling the CEO, to tell him the EPA would announce the details of its dredging plan at 11:00 a.m. the next morning. After thirty years of fighting about the issue, GE would be on the hook to take a sizable chunk of the pollutants out of the river.

At 8:00 a.m., two days before Christmas, a GE press release hit the wire, bearing the dry headline "GE Statement on Previously Discussed Fourth-Quarter 2010 Items." Regulatory approvals were coming along, it said, for GE's sale of a majority stake in NBC Universal to Comcast and that deal was expected to close in January, later than expected. The press release then said that GE had informed the EPA that it would comply with the agency's order and carry out the Hudson River dredging project's second phase.

"We engaged in intensive and constructive discussions with the EPA, and the decision reflects our discussions and many of our proposals," the statement said.

The press release closed with a note of reassurance for investors: "We expect this to resolve future uncertainty regarding Hudson dredging liabilities and to result in an after-tax charge of about $500 million in the fourth quarter. As discussed on December 14, we expect that positive items, including a favorable tax settlement, will offset this charge."

It was a smaller remedy than environmentalists and some local officials wanted. But GE would finally be digging up some of what it had dumped in the river.

PALATE CLEANSER

GENERAL ELECTRIC NEEDED to reinvent itself.

For more than a decade, GE had found itself uncomfortably defined by unwelcome comparisons and outside events — the ever-present shadow of Welch and his unmatchable financials, the embarrassing brush with insolvency during the financial crisis, the loathing of environmentalists and the sudden wallop of the disaster at Fukushima. It seemed that GE never stopped mopping up messes and patching up its image.

That job, naturally, would fall on Beth Comstock. At BBDO, GE's advertising agency, the rumor was that Comstock's ad budget had no definitive number at the bottom — it had no limit.

The marketing officer was a slim, attractive, middle-aged woman who rose up through NBC to take over the GE marketing operation. Known simply by her first name in the company, Beth Comstock had a taste for black leather biker jackets and a gift for lofty, gauzy rhetoric that belied her quiet manner. And she had long ago won the most important ear of all: Immelt's.

Comstock would continue the biggest project of the Immelt era: the reinvention of GE and its employees in the eyes of the broader public. It fell to Comstock to make the company proactive again, to get back on a forward tilt and tell the world, as it always had, why the world needed General Electric.

In Immelt's eyes, Comstock could do no wrong, not least because, in

the effulgent blasts of ad copy and social media content turned out by her massive division, neither could General Electric. Jeff Immelt hadn't been satisfied to use the same old slogans he'd inherited. GE was no longer the company that could "bring good things to light," in part because Immelt was moving the company away from consumer businesses. Instead, he and Comstock were rebranding GE as something more amorphous — not just a maker of innovative incremental improvements in jet engines and MRI machines, but a producer of innovation itself.

GE opened its wallet to make Super Bowl ads and sponsored online journalism startups. It was the main sponsor at the launch of Vox, an online news organization intended to present a new authoritative source of news and analysis. But GE would tell its own story about itself. Far from being a hidebound industrial dinosaur, GE was now fresh and exciting. The company would eventually convince multiple news outlets to declare that GE was no less than a "124-year-old startup."

In the Immelt years, it was Comstock who was most responsible for promoting the GE tagline "Imagination at Work." That phrase capped off GE's television ads and popped up in press releases and on handouts for journalists. The term, like the torrent of ad campaigns and sponsored content unleashed by Comstock's team, was intended to convey a sense of total entrepreneurial invention and exploration.

Comstock and Immelt knew that, in the new environment, it wasn't enough to just boast about GE's traditional engineering prowess and ability to make incremental improvements in the heart of heavy machinery that yielded meaningful, lucrative progress in performance as well as safety for airlines, power plants, and hospitals. They were also trying to channel for GE — a company whose longevity and solidity could make it look unexciting to markets and the public — some of the vital energy that was coursing through other parts of the economy. GE wanted to be seen in the same new and vibrant light that shone in the early 2000s on the business giants in Silicon Valley and on those with connections to the major players in tech and online commerce.

Meanwhile, Comstock's personal brand grew in tandem with the growth of GE's new marketing campaign. All sorts of new promotions and partnerships emerging from Comstock's team conveyed an experimental whimsy that was unusual for a company as hierarchical, staid, and traditional as GE. The company hurled itself into partnerships with

startups, bombarding the press with upbeat product demonstrations, special-purpose websites, and media events; it produced elaborate podcasts and even sold limited-edition hot sauce as a publicity stunt. Under Comstock, GE coproduced documentary specials with *National Geographic*'s television arm, hiring A-list directors to film short features that highlighted GE machines. The company took the normal business of corporate PR many steps further when it paid *BuzzFeed* as a sponsor of custom-built clickable web features, and it assembled its own outward-facing news site, replete with positive updates about the company's technologies, called *GE Reports*.

GE was recasting itself as not that different from the Silicon Valley hotshots. The company latched onto startups, looking to integrate some of the buzzy concepts of software development and West Coast tech culture into its traditional manufacturing operations. GE may have been older, so the message went, but it wasn't just a tired industrial company making greasy parts. It had been changing the world for decades, just like the tech companies.

"The engineers have to recognize technology isn't everything. Maybe they should work with a marketer to understand customer needs," Comstock told *Techonomy*. "We want to embrace our own entrepreneurial spirit. We were started by Thomas Edison, the Steve Jobs of his time."

It was a palate cleanser to dispel the lingering aftertaste of the financial crisis. GE was trying to show that it wasn't anything like a financial services company.

That is how Immelt wound up sharing the stage with Ben Kaufman, the ebullient young founder of a startup called Quirky. The startup was intended to be an online platform for inventors. Users submitted their designs for gadgets and machines they had invented, and the members of the website voted on which ones had real promise, offered suggestions on design tweaks, and weighed in on how the products should be named or priced. Winners got to see their inventions mass-produced and sold in stores — a transparent, outsider-led inversion of the product design, testing, and marketing process perfected by companies like GE in the twentieth century as they invented consumer appliances and convinced the public that they needed to buy them.

The innovations showcased on Quirky were modest. Its biggest hit

was something called the Pivot Power, a bendable power strip for tight spaces. Another was a $79 egg tray whose contents could be checked in real time by a smartphone app, relieving users of the burden of physically checking the fridge when planning an omelet. The tray sold modestly and was eventually pulled from production.

Kaufman was a second-time entrepreneur. He had launched Mophie, a cell-phone accessory company named after his two golden retrievers, at age eighteen. The firm pioneered the portable battery packs toted around by people to charge their phones on the run. Eventually Kaufman sold Mophie in order to focus on his next idea, which would become Quirky.

The new company had some of the goofy and freewheeling adolescent energy of some of the tech industry's most successful startups — an energy that GE hoped to exude itself. Inventions were debuted, evaluated, critiqued, and mocked at raucous weekly reviewing sessions, streamed live online.

Quirky was a lark, but it was also a serious venture. GE officials spoke of the company's consumer-informed approach as a modernizing and innovative practice that would influence GE's own appliance unit. That unit was an increasingly commoditized and low-margin business, the sort of legacy operation that Immelt, in spite of having been a top executive there, thought little of and wanted to sell off. In fact, GE had put the appliance unit up for sale before the financial crisis, ultimately pulling it back after the recession scared away potential bidders.

But now GE claimed the appliance business was providing a foundation for Quirky's amateur inventors. When one member of the web platform proposed a window air conditioning unit that could be controlled with a smartphone app, GE and Quirky teamed up to turn the idea into an actual consumer product. The result wasn't exactly the next iPhone. An existing GE window-mounted air conditioner was overlaid with Quirky software, marketed in a blur of social media hype, and sold online by Amazon. In New York City, customers could also take part in an especially trendy promotion: Uber drivers would bring Quirky employees to deliver the unit directly to their doors.

25

THE QUARRY

ADAM SMITH SPOTTED blood in the water.

It was the fall of 2012, and Smith (a pseudonym used here to protect this individual's identity) was a member of the business development team at GE's massive Power and Water business unit, the oldest and largest industrial unit in the company. "Business development" was just another of way of saying the deals team — the squad working as mergers and acquisitions bankers embedded inside the company, figuring out pieces of GE that could be sliced off and sold to generate more money than GE made from operating them. The team also kept an eye out for weaknesses in outside firms that might signal a potential acquisition of an external business for a sweet price.

Here was the blood: Alstom, the French conglomerate that competed with GE in the power turbine business and also made passenger trains, had quietly sold €350 million in new shares to private investors, just months after buying back its own stock and reporting earnings that had shown no sign that the company would need to raise money. To Smith, the sale was a clear sign of a cash crunch that one of GE's biggest rivals in the power business clearly didn't want anyone to know about.

The deals team dug down further. Alstom was heavily bloated by American standards, GE's analysis found. Compared to GE's Power business, Alstom had four times as many workers for every dollar of revenue it brought in. And yet, though it seemed to be in need of cash, it wasn't laying off workers or closing facilities. Layoffs and closures —

restructuring, in accounting speak — are cost-cutting measures taken with an eye to the long term, but because of costs like workers' severance, they require cash to complete.

It seemed to Smith that Alstom would be a prime target for the cost synergies that GE loved, namely, excess workers and factory "rooftops" that could be shed to boost profit margins. But the company was so dangerously low on cash that it couldn't do its own restructuring, not even to save itself. Buying Alstom to restructure its operations and fold it into GE's own Power unit would make sense — but only if it could be had for the right price. Any bid needed to be carefully and precisely calculated; overpaying for a limping business could hamper a buyer for years.

Smith watched his screens for days, waiting for a sell-off in Alstom shares to drive down the company's stock price. It was only a matter of time before the market saw what he had seen, and the moment that happened would be the right time to swoop in with a lowball offer to take some of the company's business units off management's hands.

But the big sell-off never came. Instead, Alstom's shares dipped, but then soon started to climb again as investors speculated about the company's plans. The market seemed to think that Alstom had raised the cash for a big move of its own. Regardless, other investors weren't seeing what Smith was seeing — a big industrial company quietly trying to manage some big internal problems.

Smith and his colleagues sent their pitch to the business development team for the entire corporation at GE headquarters in Fairfield. But corporate filed the report away for the time being. Something was obviously wrong with Alstom. Perhaps one day they would find out how bad it was.

It was April 2013, late in the evening, and Jimmy Lee was in a bar on Bourbon Street, finally getting his chance to meet the new deal guy for General Electric. Lee, the irrepressible deal fanatic of JPMorgan Chase, was the kind of fellow who could happily extemporize at length about 150 percent of the transactions that might happen in any given year on Wall Street. But GE, in recent years, had been a frustratingly sparse source of deals — and bankers' fees — for Jimmy Lee and Morgan.

GE was just the sort of sluggish and complicated colossus that Lee believed, with every fiber of his being, would benefit the most from the

skillful slicing and dicing of the nation's most august investment bank. GE was, to a banker's mind, a web of locks to be picked. Unlock them, and all-important "value" flowed out: higher returns for the rest of the company that remained, a shot at a more focused business and profits for the unit that was sold off, a surge in trading by investors eager to see something change, and a slew of fees for the bankers and lawyers who turned the notion of the deal into a reality.

It was no surprise, then, that a furiously energetic, upbeat banker like Lee could find a willing audience in a chief executive like Immelt, who was so eager to transform — and to get credit for transforming — his massive portfolio without giving up the otherworldly scale of a conglomerate in a world from which conglomerates had largely disappeared. Lee had already helped prod Immelt into selling GE's stake in NBC Universal to Comcast, a massive two-part deal. Now, in the spring of 2013, he saw a chance to bond with the new head of business development, just arrived from a posting in India, to impress on him the value that JPMorgan Chase — and in particular Jimmy Lee — were prepared to unlock for General Electric.

So James B. Lee and John Flannery, along with Geoff Beattie, a member of the board of directors, settled into the bar.

Lee's message was about playing defense. Activist investors were looking for bigger and bigger targets, he said. What defenses was GE preparing? What would they do to fend off an attempt, however unexpected, from an interloper that might try to buy into GE's stock and force management's hand about changing its portfolio?

"Look, man," Jimmy Lee said, "there's some big trends going on here and you've got to be dialed in to it."

Flannery cut Lee off. They had an idea already in motion, he told the banker. GE was ready to cut a profitable chunk of GE Capital loose, throwing a little over the side to show investors that they were serious about reducing its dependency on the financial services unit that had led to such lingering disgust.

Actually, a giant spin-off was in the making. What had been a unit of GE Capital that funded store-branded credit cards and layaway financing plans at outlets like Walmart would become its own company, with shares of stock swapped out to GE shareholders. By creating a new public company, Synchrony Financial, out of the old GE Capital, GE would

lose the money it had been making from the unit, but it would also shave off some of its financial risks.

To no one's surprise, when the move went public, Jimmy Lee and JPMorgan Chase were in on the deal.

Then, in late 2013, the word went out to the business development teams at all GE units: send your biggest industrial deal ideas to Fairfield. There, on GE's blandly imposing corporate campus, Jeff Immelt and his lieutenants were in the mood to do something big. Incremental gestures toward reorienting the company — even relatively large deals like Immelt's acquisitions in oil and gas — weren't convincing the market that GE was really serious about kicking its reliance on finance.

In the years since the financial crisis, the consensus view of GE had hardened into what had once been the wry, cynical view: America's most famous industrial company was actually more of a bank with profitable jet engine and power businesses grafted onto it. And being viewed as a bank spelled trouble for one of the most important metrics confronting Immelt and the other GE executives: the multiple.

The multiple, or price-to-earnings ratio, is a crude metric used to find the relative value of a company's stock. In its most basic form, the stock price is divided by the annual per-share earnings to determine how much investors are paying per dollar of company profit. Investors tend to pay up for earnings if they think a lot more are coming; they pay less for slower-growth companies or for those with an uncertain future. Unlike the share price, the multiple allows comparisons to similar companies.

For all of GE's growing reliance on financial services profits in the Welch years to drive its earnings up, the company had enjoyed the benefit of a double standard. It raked in cash from its lending business, just as big banks did, but investors trusted it for the stolid and secure industrial business it also was. GE had the high-efficiency profit margins of leveraged finance, but the price-to-earnings multiple on its stock price was that of a much less risky industrial conglomerate, where sales are forecasted out by years, not just quarters. Huge backlogs of heavy equipment orders and service contracts had made the company confident that even sharp economic downturns would be manageable and require no deeply painful moves. Its rich dividend in particular seemed safe from reductions.

GE's financial crisis swoon and slow recovery had changed all that. The black box of GE Capital had been acceptable to a certain cohort of big investors, sell-side analysts, and financial journalists for as long as it remained a caveat to the larger story—that of an impregnable industrial company with a knack for setting and beating targets for Wall Street. When a real financial calamity occurred, GE Capital—along with the belated panic on the part of investors about how little they understood its inner workings—had nearly taken the entire company down. Although Keith Sherin in particular had helped to stanch the bleeding in the now-legendary investor meeting with GE analysts in March 2009, during the depths of GE's crisis, and its stock price was back from its single-digit lows, there was one thing that wasn't coming back: that irksome multiple.

GE's price-to-earnings ratio now clung stubbornly to the levels of other financial services companies. To investors, the risk in GE shares was more comparable to JPMorgan Chase and Wells Fargo than to industrials like United Technologies and 3M.

For Immelt, the unchanged multiple represented a series of short-term problems. Millions of GE stock options earned as compensation by managers, executives, and board directors, both current and retired, remained underwater. Nevertheless, GE's communications department took pains to instruct reporters that Immelt, his lieutenants like Sherin and Jeff Bornstein, and the rest of the leadership team paid no attention to the share price.

This assertion was easily contradicted over cups of coffee with GE people in Midtown, and in phone calls to offices around the country. The stock price was a daily grade of Immelt's and the company's performance that could bring on a lambasting by Jim Cramer and various annoyingly well-read financial columns and blogs. Such criticism could be dismissed by Fairfield as short-termism that failed to consider the long view of the industrial company, but for scores of people critical to GE's fate, the stock price was ultimately their money.

It was one thing to abide by the principles of performance-based stock compensation when the company could take a hit over a quarter or two, get its stride back, and rise again, bringing the price back up to a range where the options that executives had earned would be worth more than they cost to cash in (the deal tacitly on offer when the options

were priced in the first place). GE's inability to climb back to the multi-ples it had once enjoyed didn't just incite grumbling from former board members and retired executives whose options were dying unexercised. It threatened to have a major impact on the recruitment and retention of the best prospects.

Still worse for Immelt, the apparent reclassification of the company by Wall Street hinted at a more dire judgment. It wasn't just that inves-tors, in assessing their risk of losing money, saw GE as more akin to a major bank than a major manufacturer. The fact that investors made that judgment was a signal of their unwillingness to believe the com-pany's near-daily protestations that it was an industrial company that happened to have built-in financial advantages. GE's investment prop-osition demanded faith in its ability to keep financial risks as firmly in check as the risks of failure in any of its industrial lines. Any jet en-gine maker would take a hit if a turbine blade broke loose and spun free from a plane in flight. A nuclear reactor meltdown would ding the stock of its manufacturer. An MRI magnet recall would incur a big charge to earnings for a healthcare equipment maker. The premise of GE's ev-ery communication to investors was that its management was so strong, and its vision of the entire conglomerate so penetrating and expansive, that its stock was a gamble worth taking. Even in the inevitable event that something somewhere went wrong, its protocols for working out of trouble would limit the amount of long-term damage and preserve its investors' money.

The security in a smaller, more stable GE Capital that Immelt now promised was that the financial services arm would be managed in just the same way: the risks it posed to a GE shareholder would not be sig-nificantly different from those posed by, say, the century-old locomo-tive division.

If Wall Street had given the postcrisis GE the industrial multiple that Immelt wanted, it would have meant that investors trusted GE to man-age all the arms of its conglomerate to the same degree of risk and se-curity. But Wall Street in 2013 wasn't offering that kind of multiple be-cause, to some extent, investors just didn't quite believe what Immelt was promising.

• • •

In late 2013, Immelt wasn't yet publicly using the term that would become shorthand for the maneuver he decided on to get GE out of its middling funk. Like a bulky forward under a basketball hoop who finds himself face to face with a double team threatening a trap, Immelt needed a way to change in two directions at once, to swing away from what wasn't working and then lunge toward what might.

He needed to pivot.

Even though the tactic he chose had failed him many times before, Immelt decided to essentially double down on it. He wanted to do a giant industrial acquisition that would ramp up the money GE earned from machines and services, opening a new stream of cash into the corporate treasury. That cash, and those earnings, could then be used to make an even bigger move that wasn't even discussed yet outside of the small groups that regularly huddled at headquarters in Fairfield and GE Capital's nerve center down the road in Norwalk. If the industrial deal worked, GE would have the financial cushion to move even further to wean itself off financial services. Instead of simply promising a smaller and more secure GE Capital, GE could make a fundamental shift away from the dependence on finance that had defined it for a generation.

But first, the Jeffs — Immelt and Bornstein — and John Flannery, the head of corporate M&A, needed a suitably big target. They needed a whale of an industrial deal, something as large as the acquisitions of Amersham and NBC Universal had been. But unlike those deals, they needed one that Wall Street would actually like.

So word went out from Fairfield to Atlanta and Chicago and Schenectady, to Cincinnati and London: send us the biggest targets in your sights.

Adam Smith and the Power team immediately thought of just such a large target — not that they thought much of the property. Still, a call for pitches was a call for pitches, and there could be much value for GE's vaunted cost-cutting machine to reap from a declining French conglomerate with assets strewn across three continents, provided they could snap it up at a buyer's price. The Power and Water division's business development team dropped their idea into the stream of proposals now flowing toward Connecticut.

What if GE bought Alstom?

26

NEW LIEUTENANT

YEARS OF FITFUL progress on reorienting GE's industrial business lines toward areas where the global economy might soon begin to grow again were doing little for the all-important stock price. Even as Immelt and his advisers privately deliberated over what big acquisitions or sales would rapidly swing the public's idea of GE's core business and strengths, the market refused to deviate from the clear signal it had flashed on and off since the financial crisis. A certain type of investor would never trust GE as long as it relied so heavily on finance to make its numbers. Finding a way out of Capital was the only way to get rid of the stock market funk that bedeviled Immelt.

He started with partial measures, not going cold turkey, but beginning to wean the company off the easy money it had grown so accustomed to. By 2013, Immelt's public forecasts about the future shape of the company were crystalizing into a vision that investors could see as well. GE would shrink its reliance on Capital while also shrinking Capital itself. Units like the consumer finance business, which backed store-brand credit cards and payment plans, would be shed and Capital's total size, measured by the value of its assets, would be allowed to shrink. GE still needed GE Capital, and assured investors that it was perfectly compatible with the other industrial units, but acknowledged that letting it grow to the equivalent of the seventh-largest bank in America was letting its heft get out of hand. Simultaneously, he said, GE could increase revenue and profitability on the industrial side of the ledger enough to

push GE's earnings toward a 70/30 split: 70 percent from the industrial side of the house, and 30 percent from finance. Then Capital could resume growing, after five years of sporadic recovery from its near-death experience in the crisis.

But this time, Immelt promised, GE would grow its financial service arm "in line" with its manufacturing businesses. GE would never get as hooked on the easy profits of lending money as it had been before the crisis hit, he reassured investors, but the company could still be run so well that it would glean all the benefits of its finance days, especially the ability to consistently nail short-term earnings targets and please the market.

Immelt was issuing a tall order to the executives who sat just beneath him in the corporate hierarchy. To make good on these promises, Immelt would have to rearrange the very highest levels of his management team.

As GE's chief financial officer, Keith Sherin had long served as a counterweight to Immelt's flights of optimistic reasoning. A tall man with the tightly cropped hair of a state trooper, Sherin rode a Harley-Davidson in the rolling hills of western Connecticut on weekends. He answered questions on conference calls in a dry drone that sometimes failed to contain his annoyance, and could be relied upon at times to pull Immelt back from where his impulses had taken him. Sherin had not been influential enough to save Immelt from creating a record of notorious overpayment in mergers and acquisitions, but he was generally viewed as a force that made Immelt consider sober realities when the CEO, the consummate salesman, could think of little but the upside.

Now Sherin's role was changing. He would take over GE Capital, becoming its CEO as the company tried to deliver on the unrealistic promise to Wall Street (which Wall Street had, in effect, demanded) to make GE Capital less crucial to GE's fate while retaining all its powers to goose the quarterly numbers. Sherin would oversee the early efforts to shed some of Capital's units and reduce the financial arm's overall size. He would also be the veteran GE leader who oversaw GE's supervision by the Federal Reserve, the discipline imposed on the company by the federal laws passed in the wake of the crisis.

In Sherin's place at Immelt's right hand in Fairfield would be Jeff Bornstein, the blunt former CFO of GE Capital who had helped the

lending businesses — and thus the company itself — weather the worst of the financial crisis and stave off insolvency. Bornstein, like Sherin, cultivated an image of fearless bluntness. He also lacked the baggage of long service as the CFO of a company whose finances remained relatively opaque and so had a certain amount of credibility with stock market analysts. They liked his unfiltered style, at least when he wasn't barking at them. Immelt viewed the job in part as a reward for Bornstein's critical work staving off the financial panic. "You can tell a lot about somebody when they're in the shitstorm," Immelt said.

If Sherin's job was best understood as redefining Capital's role, both internally and for investors, it would fall to Bornstein to take a gimlet eye to the rest of GE's sprawling business. Even if Immelt was keeping the terms of GE's success characteristically hard to pin down, one easy measure of whether his overhaul of the company was working would be whether industrial revenues and earnings rose (or didn't rise) to meet the challenge of shrinking GE Capital. Bornstein quickly saw that the company's operations, which increasingly ranged widely around the world, were riddled with clutter, duplication, waste, inefficiency, and complexity. Raised in the GE culture of Welch, with little tolerance for bullshit, he saw pent-up profit that could be gleaned from simply cutting. Releasing the synergies that would come from eliminating redundancies — that is, people GE didn't need anymore — would lead to higher earnings. As renewed attention led them to try to whip the industrial businesses into shape, Bornstein, Immelt, and the rest of the GE top brass soon settled on a name for the program they embarked on in 2013: Simplification.

The program had begun under Sherin, who worked with some of the company executives overseeing GE's foreign operations. Together they had worked out ways to cut some costs, for example, on back-office staff and customer financing. Simplification wasn't just about cost-cutting, however, as Bornstein told stock analysts, and company flacks explained to news reporters. What the company would achieve through its Simplification program was far more than just higher profits accruing from the closure of a few factories or a couple of rounds of layoffs in underperforming verticals.

In a ninety-minute call with investors and analysts that fall, Bornstein promised a deep interrogation of the business practices in GE's di-

visions. Foreign operations needed to be closer to their customers, he said, and the duplicate functions in every country that GE had set up were not necessarily needed. He questioned why GE was operating with more than five hundred separate P&Ls (profit and loss statements) for a business. The company used hundreds of different enterprise software systems to manage everything from the flow of parts to service contracts to sales data. That could all be winnowed down and reduced. Its competitors were more willing to use shared-service agreements than GE, whose different business units struck parallel deals for basic industrial services in the same markets.

On that call, Bornstein pointed to new GE facilities like an industrial park in Pune, India, designed from the ground up to be multimodal: in the same building, GE could produce components for a freight locomotive or a giant wind turbine blade, adjusting the assembly line as needed based on the flow of orders. That kind of thinking would save the company millions over time, and it reflected the efforts of its leaders to get away from the classic long-term planning model of their industrial forebears and mimic Silicon Valley instead. A leaner, simplified GE would focus less on the multiyear product development projects that had defined it for generations and try to engineer more quickly to a customer's whims — and cut its losses more quickly on what wasn't working.

In an investor meeting in November 2013, Bornstein said that Simplification also required the company to rethink corporate's role. This was a smart thing to say to investors: they were always eager to hear that money could be saved at the fortresslike Fairfield corporate headquarters, or that capital would be redirected from GE's traditionally top-heavy leadership operations and out into the businesses where products were made and sold. But Bornstein was also hinting at a GE heresy, the very question that people who wanted to stay on management's good side didn't entertain in public: what good was it, really, to continue to operate a complicated conglomerate of tangentially affiliated businesses when that model had proven to be impractical in many ways, was often unmanageable at the highest levels, and sometimes even seemed to be doomed?

"We have got $3 billion of costs at corporate that is not allocated to the businesses," Bornstein said, "and we are going through our own reimagining work now as a team in Fairfield, thinking about: What is the

role of corporate? How do we add value to the businesses? What are those things we do that enhance their competitiveness? What are those things we do that don't add a ton of value to those businesses?" GE executives were now "rethinking every dollar we spend in Fairfield," he told investors, and would continue to do so for years.

Analysts liked this kind of thing. They liked Simplification, and what it meant that GE was asking itself these questions. Just the same, some of them realized that shaking up the structure — including by cutting out the lower-level P&Ls and shedding fat in some of the far-flung industrial markets — would change the way managerial talent moved up and through GE. The company believed — as an article of unchallenged faith — that the way smart men and women graduated from one assignment to the next at GE had taught them how to become great managers.

So when Bornstein said that, in far-flung markets with revenues in the mere hundreds of millions, "you don't need a complete functional team and the infrastructure that comes with managing and reporting a P&L at that level," an analyst named Steve Winoker replied, "That was how you raised managers, though."

The company would fall back on what had worked at GE Capital, Bornstein countered. Instead of a single executive overseeing basic business functions at these smaller operations, a shared back office would handle the nuts and bolts of payroll, billing, hiring, and so on. "The way we think about it is, running a business is not about being responsible for accounts payable and payroll," Bornstein said. "It is about owning customers and markets."

All GE executives should be focused on one central, critical job: selling.

PLAYING STARTUP

ERIC RIES LOOKED the part. Black-framed eyeglasses, modestly un-tamed hair, casual dress, and an expansive way of gesticulating as he made his points about how the company of the future would run. Ries was the author of a runaway business book hit, *The Lean Startup,* which professed to explain why the companies that had best captured the public imagination in the second decade of the 2000s had succeeded and how bigger, slower, older companies might emulate them to succeed as well. Ries parlayed the success of his book into speaking engagements and conference appearances, did consulting, wrote more books, and prosely-tized "lean principles" to the leaders of big corporations. Through Beth Comstock, Eric Ries came to the attention of Jeff Immelt.

The lean startup methodology borrowed in part from lean manufac-turing, which GE and other big American manufacturers had already begun to rely upon. Lean manufacturing slimmed down the traditional business model of maintaining inventory warehouses and stocking up factories with supply chain parts in favor of a constantly running de-mand engine that took in pieces and churned out assembled machines "just in time" — that is, as they were needed. Ries applied the lean prin-ciples that executives of Immelt's age had been using to wring new levels of efficiency out of manufacturing companies since the 1980s to the ven-ture capital–funded startups that dotted the economic landscape of the 2010s. Ries argued that new firms as well as more established companies needed to act like the most successful software startups. They should de-

sign fluid responses to consumer demand, iterate and reiterate product designs, and be unafraid to go to market with what wasn't yet perfect or to walk away from what had been judged a failure.

This was, in short, not the way GE worked. Designers and engineers in the appliance business planned months in advance, developing diverse lines of products to hit every market niche, in anticipation several seasons ahead of retailers' holiday sales. Even less accustomed to such a freewheeling operation were the Power, Aviation, and Healthcare businesses. Their products were incredibly complex machines whose success or failure was measured in tiny increments and by thousands of hours of service without a breakdown. The stakes of an industrial update could not be higher: unlike a software patch pushed out before its time, the consequences of failure could be dire, up to and including death.

Still, executives like Comstock, Immelt, and some of their colleagues were convinced that such arguments arose from the fading throes of resistance to an inevitable adaptation. GE wouldn't pump out jet engines that would seize up like bad apps midflight. But the physical, manufactured world appeared to be experiencing a sea change not so far removed from what was already under way in the economy of software and cloud computing. The promise of 3-D printing — or "additive manufacturing" in GE's wonkier preferred phrasing — shimmered like a mirage in a future not that far off. A radically new method of engineering metal parts could solve problems of weight and efficiency that had been intractable for decades, especially in the all-important design of jet engines.

The technology was far from perfected, but GE was starting to experiment — for example, by shaping the first rough parts of its products entirely on additive printing bays, then having human workers finish them by hand. The printers now could produce a jet engine fuel nozzle that had previously been made of twenty parts out of a single piece of metal that was significantly lighter. It did feel as if a higher-tech industry's innovations were beginning to manifest all around the modern industrial world, and Immelt was convinced — not only at the urging of self-styled thinkers like Ries but also by his own instinct to make heady bets — that GE's path forward should be to surge ahead as a leader in technology. The strategy would follow the story line that the company would tell. GE imagined itself a leader in innovation. And so innovate it would.

And Immelt used it to distinguish himself from the management mumbo jumbo and financial engineering that so many associated with GE, saying in 2016 that "additive manufacturing makes a shitload more sense than Six Sigma did."

The hope was also that embracing the language of innovation would prove to be a self-fulfilling promise. If GE talked like one of the buzzy tech unicorns dominating the business press and seemingly taking over the nation's economy, maybe it in turn would also be covered like one. Immelt was frustrated, he acknowledged in occasional asides to GE executives and board members, at how intractable the narrative about his company and himself had become. He and his top aides could point to any number of exogenous shocks to explain some of the company's sluggish performance under his tenure: the attacks of September 11 and the ensuing recession; the financial meltdown and the years of economic uncertainty it wrought; or the political distractions of the postcrisis years, including the rise of an isolationist and anti-establishment American right wing that had increasingly hamstrung GE's symbiotic and profitable relationship with the federal government.

But ultimately, though Immelt's flacks claimed that he never considered it, it all came down to the price of a General Electric share. Although it was widely acknowledged that GE stock was never going back to what it had been under Welch at the peak, certainly no one was happy that it remained so middling. The recently retired executives and directors muttered about their underwater stock options. The business TV commentators invoked it in the never-ending comparison contest between Jack and Jeff—a contest Immelt was always losing. While Immelt adhered to his GE training to fixate on shareholder return and the quick beats that would drive shares higher, he and others in the company wondered in frustration why the world couldn't just focus, for a minute, on something else. There was so much to brag about in the company—the sheer grandeur and force of the things it made, the legacy (however much it had to be spun) that stretched back to the most famous of American inventors, the way the company remained deeply embedded in American communities, still employing hundreds of thousands of people not only around the world but across the nation.

In spotlighting these impressive features of the company, Immelt and the GE message machine wove strands of rhetoric that imbued its

underlying businesses with a greater purpose, an association with ge-
nius. The turbines and engines and the factories in which they were
made were not just impressive, but "brilliant." The company's business
strength lay not so much in the designing, building, and selling of intri-
cate machines, but in the feats of imagination that conjured them into
being. Her years in the entertainment business had taught Comstock to
always return to themes of inquiry and adventure — to keep the mission
focused on associating the familiar GE meatball logo with forays into
the unknown, with discovery. A BBDO advertisement, for instance, fea-
tured an ungainly-looking feathered beast being coaxed into revealing
its plumage: a representation of nothing less than GE's unique ability to
cultivate The Idea itself.

GE aggressively pushed the business press to cover it in the same op-
timistic tone and to suspend disbelief occasionally in the same way it
did for companies from the purported new economy. GE pushed the
story of itself as a company every bit as innovative as Tesla Motors, as
attuned to changing consumer attitudes as Apple, not just in traditional
news outlets and on cable channels but in the flourishing ranks of new
social media–directed news and analysis platforms, like *BuzzFeed*, Ins-
tagram, and *Bloomberg*'s TicToc — journalism as short and shallow as
a post or a tweet.

The daily deluge of upbeat, speculative granules about technologi-
cal breakthroughs — like the musings by Tesla CEO Elon Musk about
high-speed travel by "hyperloop" — lacked or downplayed any of the
inconvenient context of a traditional news report, particularly caveats
about the obstacles between the present and the unicorns' rosy visions
of the future. GE and Jeff Immelt wanted in on that. Their language was
changing too. Like the ride-hailing company Lyft essentially "invent-
ing" the concept of the bus, GE was no longer content to boast about
running efficiently, or well, or even better than any other company, but
only about changing everything. Its system for performance reviews, the
structure of its annual shareholder report, its approach to digital tech-
nology — in all these areas and others, GE announced itself as revolu-
tionary. The company said it operated just like the freewheeling venture
capital–funded companies churning down from Sand Hill Road; lead-
ing with slick design and the promise of widespread adoption, GE's PR,

like the startups,' was less focused on how its business units were going to turn regular, dependable profits.

With its obliging corporate messaging engine revved up, GE went into the content business. The company drones swooped through turbine halls, capturing Instagram footage of its brilliant machines. In Washington, the company lured reporters into demonstration booths to learn about how FastWorks, its implementation of the gospel of Eric Ries, would upend the design of home appliances, and how GE, partnering with Local Motors, would reinvent the business of making cars through 3-D printing. To celebrate the forty-fifth anniversary of the Apollo 11 moon landing, GE's PR team paired with a shoemaker and website called JackThreads to make a sneaker based on the astronauts' moon boots, a reference to the fact that a type of silicone used in the boots worn on the mission had been designed by GE. The company finagled a photo of Buzz Aldrin wearing the shoes and arranged write-ups in places like *Fashionista*, *Adweek*, and *The Verge*, outlets that a previous generation of corporate executives wouldn't have recognized, much less cared about.

By March of 2016, after GE had spent years pressing the case, *Bloomberg Businessweek* obliged with the headline of Jeff Immelt's dreams: "How GE Exorcised the Ghost of Jack Welch to Become a 124-Year-Old Startup."

28

COWBOYS INTO FARMERS

A s GE COURTED new media attention and sought the public adoration it so envied in the startup economy of Silicon Valley, real life didn't get any easier. Not with the Federal Reserve sitting in the offices of GE Capital.

No one had ever had to worry about bank supervision in the good old days of GE Capital. Back then, the business had proudly run on something like an honor code. Sales teams chased deals, won approval on the strength of their pitches to higher-ups, and either managed to make money for the company or suffered the results of an investment gone south, such as a docked bonus or a missed promotion.

But this model was anathema to the postcrisis New York Fed. Where had the company actually written down, the bank supervisors wanted to know, the absolute criteria for extending a line of credit? (They hadn't.) Where was the risk analysis of the loans on the Capital balance sheet? Why did it have such rosy projections about which customers would pay their money back? (Capital was supremely confident in its ability to lend and foreclose without losses like the banks had suffered.) And what the hell had GE Capital been doing buying a Brazilian oil-drilling rig? (Long story.)

The teams demanding answers to these questions in the spring of 2013 reported to Caroline Frawley, the head of a New York Fed unit that had been set loose to pore through GE Capital's books after the post–financial crisis Dodd-Frank reforms were enacted. This scrutiny

sounded reasonable enough to majorities in Congress, who, having wit-
nessed GE's near-collapse in the crisis, thought it made sense to have the
same regulators who were keeping a closer eye on the assets and balance
sheets of big banks also looking at the same information from GE Capi-
tal. Capital, in the months leading up to the crash, had been effectively
the seventh-largest bank in the country, but its operations had been al-
most impenetrable. It was a "black box," everyone had said with a shrug
— until the black box had almost dragged the company down for good.

Bank regulators and GE Capital's proud core of deal-makers didn't
see balance sheets in the same way. And they really didn't like each
other. What Capital viewed as flexibility, and a culture of savvy deal-
making, professional bank supervisors were apt to see as the absence
of any meaningful rules at all. And GE Capital's books were, to those
experienced with the protocols of modern banks, a riot of tangential
and unmatched risks. GE Capital's deals — some of them entered into
only to aid the industrial business units, and many of them just specu-
lative bets — were so wide-ranging and hard to value that the exchanges
between the Fed teams and Capital executives often turned combative.
Capital had purchased the oil-drilling rig as part of a bet on an increase
in super-deep oil exploration off the coast of Brazil (a bet the broader
oil market would soon turn against). It was one of the curious assets on
the Capital books that had gone south, to the surprise of the Fed team.
The legal entity that formally owned the Peregrine I, a "dynamically po-
sitioned drilling vehicle," had filed for bankruptcy two years earlier.

The rigorous protocols of the banking industry also bore little re-
semblance to the customs that prevailed at GE Capital. Like the parent
company, GE Capital's system for evaluating and monitoring deals was
rooted in personal accountability more than enterprise risk.

A GE compliance official hired to help deal with the incoming at-
tention from the Fed supervisors asked what Capital's standard proto-
col for extending a new line of credit required. What chain of command
signed off on the deal? What committee reviewed it? These were ques-
tions that any bank would have answered easily. But at Capital, the new
official eventually learned, there was no concrete protocol. The guy who
brought in the deal was responsible for it, and his bonus was on the
line if the loan didn't work. There was no hard and fast rule, however,
about how basic issuances of credit should go. GE defended this model

as the essence of executive accountability. And investors were getting what they always said they wanted: deal-makers' pay being exposed to their success in achieving the goals of the business.

But to Fed supervisors, this was a recipe for sending a system out of control. GE's protocol, or lack of it, wasn't just unusual practice. It was dangerous, both to the company and to the very system the Fed was trying to protect.

"I am not going to let that woman tell me how to run my company."

Jeff Immelt's voice was compressed by anger into a hiss. It was 2014, and Immelt and GE's leadership were boiling over with resentment about one woman in particular: Frawley, whose team of supervisors from the Federal Reserve were clambering all over GE Capital and its inner workings, demanding reams of information about the company's deals, procedures, and debts.

Even worse, for Immelt, was Frawley's insistence on using every power the law provided her. She was a hard cop who followed the letter of every statute and regulation. That included Fed officials being able to sit in on meetings of GE's board of directors and listen to discussions about the company's financial exposure. The Fed supervisors were also meeting with the company's directors outside the confines of board meetings, without Immelt or his lieutenants present, and sometimes one-on-one with the directors, urging each one to take a harder line with Immelt. They felt that directors needed to question him more aggressively about the amount of risk that remained on GE's books half a decade after the company had almost collapsed under Capital's weight.

For a CEO accustomed to dictating the content and tenor of his board's deliberations, this felt like intrusive meddling, if not an outright assault on the company. It was also costing GE a hell of a lot of money. Privately GE executives threw around the unprovable estimate of $1 billion a year in new costs to meet the compliance requirements dictated to them by the Fed's bank examiner teams. What had once been a team of near-autonomous deal-makers who thought of themselves as cowboys more than farmers, in one executive's turn of phrase, was now a fenced-in operation. The vast portfolios of one-offs, "reasonable expansions" into new markets, and partner plays — Capital financing exotic projects far afield to loosen the soil for some of GE's industrial businesses to sow

the seeds of future sales — were all now constantly scrutinized by the Fed.

GE Capital had always considered the quirky diversity of its books its strength. Certainly in the "mid market," where the company had long thrived, GE deal-makers felt that they were deeply familiar with niches that banks ignored and that, thanks to the corporate parent's massive balance sheet, they could compete by the narrowest margins to win business. Hence the outstanding loans to snowmobile makers in South Dakota and potato chip makers in Ohio. Capital had armies of deal-makers who were not so much bankers as door-to-door salesmen hawking money. In those mid-market swaths of the company, there were few MBAs who boasted a Harvard pedigree and wore blue suits.

"Our folks come across to people as GE people," a Capital executive told one of us in 2015. "We don't wear suspenders. We don't have MBAs, for the most part. We don't come across as sort of a Wall Street, big bank kind of culture."

What they did have was deep knowledge of the assets that backed the loans they were writing and a better track record than the big banks at completing what Capital folks called "workouts" — foreclosing and seizing assets when the borrowers couldn't pay. Capital prided itself on knowing the market for the railroad tank cars that it accepted as collateral, or for the manure-spreaders that backed loans to farmers in the Midwest.

True, the unit had gone crazy for leveraged buyouts in the 1980s and '90s, only to see many of the targeted businesses go belly up in the recession. As one former GE Capital executive quipped, that had been an experiment in private equity by accident. But they had managed to keep businesses standing, turned them profitable, and finally resold them, sometimes years later. GE Capital people were sure of themselves, even as their operations got larger and more complex and broadened into areas of finance where the bigger players played and the risks were potentially catastrophic — as in their growing portfolios of real estate and the lingering remains of the insurance business. Capital people, imbued with tribal pride in their unit, insisted that they knew what they were doing.

The Fed treated them like idiots. Or at least that's what it felt like to many of GE's executives, Immelt included. Some of his directors felt the

same way. One marveled in 2014 at the tone of the written exchanges between Frawley and her team and Keith Sherin and his top deputies, which the audit committee reviewed in its role as oversight over Capital. Why, the director asked, did the Fed team always have to sound so mean?

But Frawley's skepticism of GE's internal controls wasn't undeserved. Even now, years past the crisis, Capital remained a black box even to those who were paid to understand the company intimately, like sell-side stock market analysts and even some GE executives. The company's remaining risks were unfathomably deep, some prescient investors felt. And the hubris that characterized decision-making elsewhere in the company could be just as prevalent at Capital.

One former GE executive recalled attending a presentation at Crotonville just months after the company had barely staved off disaster in the depths of the crisis. One of the speakers that day was Mike Neal, still the CEO of GE Capital. Neal embodied GE Capital. He had captained the second half of its gargantuan expansion in size and importance to the company, warmly rallying his troops in his thick Georgia drawl. He stood out like a blue collar in a business full of white ones, one of his admirers said. Under Neal, Capital had somehow avoided disaster and managed to close the books on 2009 in the black, a fact that the company would brag about for years afterward.

Neal's presentation that day on what GE had learned from its near-death experience amounted to a fatalistic shrug. He showed no outward sign of self-examination or contrition about the risks the financial unit had taken on, the dangerous excesses it had driven in a quest for earnings, or the way Capital had gambled the company, knowingly or not, on a roulette wheel of short-term commercial paper and nearly lost. Neal's analysis was macro. He spoke about what had bedeviled the larger financial system and the signs in the market they'd have to do a better job of reading next time, without acknowledging that GE should never have been in certain markets in the first place. GE had been hit, Neal seemed to be saying, by the unforeseeable, by a one-hundred-year perfect storm. And they had weathered it.

The former executive listening to Neal's version of events in Crotonville that day squirmed in his seat. *My God*, he thought to himself. *You have learned nothing.*

29

PROJECT HUBBLE

As the spring of 2013 unfolded, Mike Neal had waited as usual for his annual consultation with Immelt. Although the documents on file at HR might have suggested otherwise, the business unit CEOs served at the pleasure of the boss, and Immelt's ability to nudge them aside was always unquestioned. As it turned out, Neal was not allowed to stick around to see what would happen to his beloved GE Capital.

Big changes were coming for Capital. The market would not yield to Immelt's desire for a revaluation. As long as Wall Street believed that America's great industrial company was in fact a massive bank with a couple of turbine businesses bolted on the sides, nothing about Jeff Immelt's reputation or the company's stock price would change meaningfully.

It wasn't just the stock multiple's resistance to change that bedeviled GE Capital. The unit was discovering that life was different in the aftermath of the crisis, and not just because Caroline Frawley's team was patrolling the halls and listening in on board meetings. The essential element of GE Capital, from the Welch revolution on, had been the cost of money. They had borrowed low, thanks to GE's sterling credit rating, and lent at aggressive rates, undercutting banks for narrow bands of customers that their competitors didn't fully appreciate. Whereas banks with hoards of depositors' cash could always find the lowest cost of capital to lend out — nothing was cheaper than a dollar already in hand — using its huge stores of commercial paper had enabled GE to pick its

spots and find deals, sometimes holding on to more of its booked loans than a bank would. The crisis had upended all that.

Now, in the aftermath, GE's reliance on commercial paper to fund its deals and lending had become untenable. And at the same time the re-trenched banks, flush with the greater stores of deposits and capital that regulators insisted they hold on to, were chasing the deals that GE Capital had once had the lead on. "There's just so much money now chasing new business," the Capital executive said. To chase those deals in turn, GE was squeezing its profit margins.

In one sense, this was just the story of finance in the postcrisis period: figuring what could and couldn't make money in the new regime of rules put in place after an orgy of profit madness had nearly crushed the global financial system. But the wrinkle in the story at General Electric was the new relevance of a question that not even the near-destruction of the company had prompted, but that seemed germane now that GE Capital's margins were flagging and its usefulness in managing corporate earnings was waning: was GE Capital really worth the trouble anymore?

A newly hired General Electric executive was waiting by the gatehouse at the entrance to the corporate campus in Fairfield one morning in 2013. Suddenly, one of the guards looked up at a colleague in a panic and began shouting, "Gate! Gate!" The second man leaped forward and landed hand-first on the button that opened an entrance gate just in time to lift the bar out of the way of a car that roared past up the drive, never braking. The new hire thought it looked like a late-model Porsche. Behind the wheel, unmistakably, was Jeff Bornstein.

Bornstein was a hard-charging guy, short with a close-shaven, balding head and the presence of an enforcer. The GE press staff confirmed on background that it was nicotine gum he was so often chomping on when he made the rounds, glad-handing the investors, parrying the sell-side guys, gracing the reporters with an off-the-record quip or two. They also confirmed that he was a shark fisherman, though Bornstein himself would clarify that it was tuna he hunted. Despite his reputation, Bornstein was charming. He could be condescending and even harsh when he dressed down subordinates, but he had a warm cackle and disarming

candor that many in the company's upper echelons lacked, especially his predecessor, the automatonic Keith Sherin.

Early in Bornstein's tenure as CFO, one GE director remarked that he wasn't the GE type, but in a good way. This person meant that Bornstein spoke candidly about things that had to change and was unaffected by the nostalgia for GE's traditions and long-standing business lines.

Having reached the job of CFO, Bornstein now sat at the apex of an obscure internal hierarchy of financial operatives who were interspersed through the company's scores of companies and divisions.

By early 2014, the meetings had begun — always in secret, always in small groups. These informal ad hoc gatherings — which included Bornstein, Denniston, Immelt, Sherin, and a handful of others — were considering the unthinkable.

What if they could find a way out of GE Capital?

Such thinking had been bolstered by the success of spinning off part of GE's consumer finance business as Synchrony in 2014. The first plan explored by the company's teams, dubbed Project Beacon, was the possibility of slicing GE Capital into a separate independent company. This analysis showed that the move would entail massive tax costs and other risks. A second approach was developed to simply sell off the major pieces of the division; this alternative, dubbed Project Hubble, would still have large tax costs and bring its own uncertainties. The final decision would come later, but for now GE was heading in the direction of Hubble.

The complications were enormous. First, of course, they needed to find buyers for huge, diverse assets in the GE Capital balance sheet — office buildings and corporate parks, rail tank car fleets, rafts of loans to fast-food franchisees and farmers. They would need to resolve the taxes, including the disposition of billions in deferred losses, which now were accounted as tax assets that GE could use to boost its earnings. Above all there was the consideration of earnings. The successful sale of GE Capital's parts would generate big onetime gains, the GE executives thought, especially if they were all sold off quickly. Those gains could be reinvested, either in industrial businesses or, preferably, in stock buybacks, which would mop up outstanding shares and boost the price.

But GE would also need to prepare for a vastly different future in which the parent company would be operating without the reliable, regular dividend of earnings flowing up from Capital into the corporate treasury. Could GE reliably hit its earnings targets without the tools it had used at corporate to smooth out rough quarters and boost its paper profits when needed?

The leaders of the company were quietly considering a kind of disarmament. They wanted to see if they could finally shed their tremendous, but troublesome, financial weapon by laying it down entirely. They considered the question with the characteristic self-assurance of GE men. Even without the use of Capital — which some likened to a steroid that boosted the company's performance artificially — they were certain that they could generate huge profits, pay the unusually rich dividend that kept shareholders happy, and squeeze greater profit out of the aging manufacturing base of jet engines, power turbines, and MRI machines. They would keep going global — expanding abroad into new markets in search of the growing international middle class whose demands for a higher standard of living underpinned GE's commercial strategy. They would gobble up market share in everything from power turbines to medical devices. They would be rewarded by Wall Street for moving on from finance. They would tell the world a story about innovation and excellence, and they would be believed. As the spring of 2015 arrived, it was clear that GE was going with Project Hubble, which was still a secret kept from everyone except the highest echelon of GE executives and a small team of its most trusted bankers, including Jimmy Lee.

GE was getting ready to sell Capital.

30

THE CHOP SHOP

A SERIES OF PHONE calls and texts had the GE communications shop worried. It was April 2015, and they were hours away from unveiling the fruits of Project Hubble, a multifaceted deal to unload a huge portfolio of real estate and mortgage assets to Blackstone. In effectively exiting the troublesome real estate business, GE was announcing the broad outlines of what would come to be known as the "Capital exit plan."

Nearly all of GE Capital was going up for sale. Placing a premium on speed, the company would offer to sell its lending operations in chunks. Big banks and private equity funds would snap up the obscure pieces of the lending operation, taking over loans to small businesses, machine middlemen, and leveraged buyout shops. The plan would call for GE to hang on to GECAS, the jewel in Capital's crown; that unit owned and leased a massive fleet of planes and jet engines, reaping profits with every launch of a commercial airliner into the air. It would also hang on to the stubs that facilitated the operations of the industrial businesses, such as those that lent for the development of new power installations or financed the purchase of healthcare equipment.

GE would also be left with what amounted to the unwanted detritus in the basement of GE Capital — the odds and ends of deals gone sideways or bets that hadn't panned out over the decades of aggressively pursuing returns with borrowed funds. (To the surprise of some observers, GE's investor slides would include a graphic showing a sliver of "insurance" among the assets that would be left behind in the Capital exit. In-

surance was a business GE was supposed to have left behind a decade earlier.)

If top management pulled it off, the announcement would put a charge into GE's stock, sending it soaring on hopes that the lumbering corporation was still capable of moving quickly and shedding the dependence on lending that its most critical investors simply hadn't trusted.

Like most major GE news, the announcement was being carefully stage-managed. The news would break simultaneously on CNBC and in the *New York Times,* which had been supplied with the news under an embargo. But now reporters from the *Wall Street Journal* were sniffing around, sensing a major deal afoot. Reporters who covered real estate had gotten wind of a large and mysterious transaction under way. Colleagues on the deals team were bombarding banker sources, trying to gauge the scale and price of what was on the block. Corporate beat reporters called sources of their own, grilling GE executives and members of the board about what the company was up to and asking why the deal was shrouded in an unusual level of secrecy.

Company flacks called reporters at the *Wall Street Journal* to offer a major story if the paper would agree to an embargo. Already close to a scoop, the reporters and editors conferred and decided, with the competition tightening, to decline.

The first piece of the news appeared in a 145-word story on the *Dow Jones Newswires* in the early afternoon of April 9. "General Electric Co. is close to selling all or part of its giant real estate holdings, people familiar with the matter said, as the conglomerate moves more aggressively to shrink its finance business," the *Journal* reported, leading with the news of one of the largest property deals ever as its reporters raced to confirm what else was coming. GE's stock began to rise that afternoon as investors seized on the sign that Immelt might at last be ready to make a serious move away from finance. The real estate move alone would have been a huge story, but it was clear to reporters and editors in New York that something much larger was afoot as well. An all-nighter ensued as reporters called and texted tight-lipped sources who were sworn to secrecy, pleading and badgering for information about the deal.

By the early morning hours of April 10, the boyish and inexhaustible corporate editor for the *Wall Street Journal,* Drew Dowell, was at home in New Jersey, conferring with reporters as his finger hovered over his

computer mouse, ready to hit Send. A key source had come through for veteran *Journal* reporter Joann S. Lublin, a newsroom fixture for more than four decades whose tenacity in wearing down corporate titans who dared to resist her inquiries about hirings, firings, coups, and deals was legendary.

Dowell weighed the sourcing, the strength of the confirmations, the gravity of the news the *Journal* was about to report, and the fierce urge to beat the competition — especially those he could be sure had the whole story straight from the company, tied up in a bow and waiting for the appointed hour to publish. He hit Send.

The *Journal* headlines hit the wire at 6:25 a.m., five minutes before the GE press release: "GE Prepared to Exit the Bulk of GE Capital — Sources."

GE's announcement landed with the bang that Immelt, Bornstein, Sherin, and the board had hoped. The stock shot upward, and commentators praised GE's move. The news coverage noted that Immelt was putting his own unmistakable stamp on the company, utterly altering the vehicle he'd been bequeathed by Welch. And there was more good news to come, the company said.

In the public announcement of the exit plan, GE said that the asset sales would be completed over twenty-four months, obviously setting a target for GE's executives to hit and surpass. The focus, the company insisted, would be on divesting these assets quickly. The logic behind that strategy was twofold. First, a drawn-out period of separation could depress morale and distract from normal business. Indeed, even the relatively speedy process that ensued had its awkward moments. For instance, a temporary wall was erected to separate one of the units slated to be sold but not yet disposed of from the rest of the Capital office suite in Norwalk, Connecticut.

Second, many of the business units were really only as valuable as their people. Holding the value of Antares, the internal unit that provided financing for middle-market private equity deal-makers, required selling it off intact, and fast, before its leading executives could be recruited away by competitors. And those competitors were already circling, their interest especially directed at poaching Capital executives from units whose ultimate destinations were unclear.

Suddenly executives had little prospect of a career-long stay at GE

unless they found a new perch back in the parent company. And there was no security in waiting for their business unit to be swallowed up by a large bank or fund, which would have its own arcane system of seniority and command as well as the classic tool for ensuring the profitability of a new acquisition: laying off redundant employees.

So GE promised swift execution of its sales, and Sherin began whipping sell teams to get buyers interested, set up data rooms, and begin the hard work of valuing Capital's often amorphous P&Ls for the arriving teams of bankers.

Some found the timing curious. Was GE selling at the right time to get the most value for these businesses it had built over decades? GE didn't care — it wanted out. It saw the opportunity and needed to take it. Otherwise, as Immelt told people, they ran the risk of another wave of contagion spreading into the financial markets if they waited for something better. That could be the end of GE. It was better now to sell on their terms.

GE said that its decision to sell off Capital would focus and energize the already promised reorientation to its industrial operations, which would provide 90 percent of GE's earnings by 2018, the company said in its announcement. It also admitted that the change wouldn't come cheap. GE would take an after-tax charge of $16 billion, mostly noncash, to account for the cost of unwinding Capital's complex tax schemes and repatriating the billions in cash it held in operations overseas. But the company would shower its existing investor base in profits from the deals. The Capital asset sales were projected to yield $35 billion in cash to the parent company, the press release noted, adding that the board had authorized a new round of stock buybacks, up to $50 billion.

GE's move was extremely self-assured. It projected the image of a company that had aggressively used its financial wing for a generation to manage its earnings and now was telling the world, backed by its Wall Street reputation, that it could shed that piece of itself and not just fail to suffer but actually perform better than it had in a decade.

Some executives, even some within GE Capital, worried about GE cutting off a source of high-quality earnings: payments on leases, loans, and buildings came in hard cash and flowed up through the internal dividend to the corporate office. GE had plenty of earnings, but many of them didn't represent cash payments into the company coffers. When

the service businesses stretched, they created profits on paper, and so the quarterly search for actual cash could grow desperate in the company's lean times. Turning off a reliable cash source, these executives thought, was not something to be done cavalierly. As one former senior executive put it, GE would be naked to the world. But it wasn't in the company's DNA to doubt itself. Such self-questioning was akin to disloyalty in the eyes of leaders like Immelt.

One morning at the GE Capital headquarters off the Merritt Parkway in Norwalk, Sherin walked into an office where several executives who reported to him were gathered. The exit plan was still a secret, but top-level officials at Capital were beginning to be brought in on the plan.

"We're going to de-SIFI," the GE Capital CEO said. That is, GE was planning to shed the Systemically Important Financial Institution categorization that had brought it under the scrutiny of the Fed by selling off most of its finance operation. SIFI status was the regulatory scarlet letter that followed the financial crisis: the designation was applied to the largest banks and financial institutions, those whose failure could trigger another meltdown. The only way to lose the label was to shrink.

"What are we going to do about the cash?" one executive in the room said, wondering how the company would pay its dividend or make the massive investments that the industrial businesses required.

Sherin shrugged. "It's just work," he said, repeating a phrase that would become a refrain in the following months filled with single-minded focus on pursuing negotiations, finding buyers, closing deals, and shutting down the bulk of Capital. There was no plan B for replacing the cash that GE would need to keep operating as it had. Plan B was to find it somehow, to figure it out.

In fact, Immelt, at least, thought he knew of one source of the cash they all craved. It had been set in motion months earlier, before any but a few executives had any inkling that the company planned to get rid of GE Capital. When the word had gone out in early 2014 from Fairfield to the business development teams embedded in the headquarters of all GE industrial business lines to send in their biggest M&A proposals, the bankers within the company all recognized the opportunity immediately. Jeff Immelt wanted to strike a really big deal.

31

BIG BETS

THE NUMBER $30 loomed.

GE's stock price and its miserable performance were a constant cloud over Immelt's head.

He had transformed much of the company he inherited. He'd sold vestigial and beloved business units. He'd plunged deeper into entertainment, vastly expanded the global footprint. Delivering his perennial GE halftime orations — the football player's rhetoric translated into exhortations about "winning" and "blocking and tackling" to hundreds of midlevel executives across the conglomerate — he was leaving his mark on the company.

And yet GE had been nearly destroyed by the habit that Jack Welch had developed and Jeff Immelt couldn't kick. Finance had nearly brought them down. They all owned their reliance on it, even those who, like Immelt himself, didn't fully appreciate the risks it exposed them to. And now that the financial crisis had passed, now that they had somehow survived it, GE Capital lay like a brick on the company's spine.

Each infuriating news report about GE's latest quarterly earnings contained some version of the same line, one the communications staff tried endlessly to argue out of articles, without success. It was a refrain that drifted like sour vapor among scores of current and retired executives whose compensation was tied up in stock options that were largely worthless.

It had been six years since a share of General Electric stock had fetched

$30 on an open exchange, the stories all pointed out, much to the CEO's frustration. Despite all Immelt's changes and the company's miraculous survival of the crisis, the stock persistently traded, not at the valuation of an industrial colossus, but like a mortgage company or a bank.

What the share price augured for Immelt wasn't a Welch-like legacy. What it hinted at for a CEO who stood in constant, uneasy comparison to his legendary predecessor was a legacy that wasn't even good. And though Immelt planned to stay in his position for years, time was running out to change the overarching narrative of his tenure.

Immelt knew that changing his legacy would require even more momentous changes in the architecture of the company. He needed to make moves that would finally impress upon Wall Street that he had found a way to lead the old General Electric into a new economic paradigm — causing its share price to respond accordingly.

Clearly, he'd have to lessen GE's dependence on finance, but he would also need something, in one of his favorite words, "additive." A master stroke to convince his critics that Jeff Immelt, just like Jack Welch, had the foresight and gumption to steer GE toward ever-greater profits. Immelt needed a big deal.

He would find his quarry in France. In the meantime, Immelt and GE continued doing what the company had done so readily under Welch: moving in and out of businesses, to the rhythm of a changing economy, as a protean corporation with an unchanging industrial core and an outer shell rigorously updated with the market trends and buzzwords of the moment.

Out in California, Bill Ruh was trying to convince a skeptical industry to believe a bold proposition: General Electric was a software company.

Ruh looked the part of the midcareer manager he was: with thinning brown hair and a sizable paunch, he was quick with a friendly chuckle. Like Immelt, he had a facility for the latest of the company's buzzwords. At the moment that message was a single word: digital.

Jeff Immelt had seen the future of industrial companies: it was in software and hard-core computing. His vision was one that other industrial company leaders, including those who laughed at GE initially, would increasingly come to share. Immelt saw that, as technological innovation shrank the size and cost of digital sensors, they would be embedded

in more and more machinery of every conceivable type. Huge streams of data could then be harvested, not just from smartphones and WiFi-enabled thermostats, but from the massive machines at the center of the developed economies around the world: gas-fired power turbines, jet engines on passenger aircraft, MRI machines and sonograms, the rumbling diesel-electric locomotives hauling freight across the American plains or down the core of the Indian subcontinent from Delhi to Bangalore.

There was support for Immelt's foray into software in the traditionally gritty industrial world. Marc Andreessen famously wrote in 2011 that "software is eating the world," meaning that it was transforming and disrupting businesses and sectors throughout the economy. But the widespread innovation, he noted, wouldn't be as destructive for certain companies. "In some industries, particularly those with a heavy real-world component such as oil and gas, the software revolution is primarily an opportunity for incumbents," he wrote. "Over the next 10 years, the battles between incumbents and software-powered insurgents will be epic."

GE was determined to win that battle. To Immelt, his board of directors, and the executives who shared his vision, GE stood the greatest chance of harnessing these streams of industrial data. The company would know best which data to gather and how to interpret their meaning because they knew the machines; they had designed and built and repaired them for decades, after all. Reams of digital information were already being generated by the massive fleet of GE-designed jet engines that roared down runways each day. Working backward through the numbers, engineers empowered with the right digital tools could begin to learn new things from their own machines, Immelt said. "Big data" would help GE better predict when critical parts in engines and turbines were likely to wear out and break.

GE would then repackage that knowledge and sell it to customers. Service contracts could more reliably guarantee customers like airlines and utilities a minimum number of hours of smooth operation. Better planning for maintenance would flatten out and hold down labor costs for GE. And the digital model would provide critical "stickiness" — a guarantee against untold future innovations that might allow third-party businesses to swoop in and poach service contracts on GE-built

machinery. That potential, which was already a problem in lower-tech industrial businesses like elevators and building systems, was a distant but real threat for all industrial businesses. It was in servicing industrial machinery — not in selling it — that all the profit lay.

Immelt would later tell the press that this revelation came to him in the aftermath of the financial crisis, on a visit to a GE jet engine factory where he heard engineers talking about the sensors that would be fitted onto the most modern engines. Where, the CEO wondered, was all that data going?

By 2011, GE had hired Ruh from Cisco and begun building the campus at San Ramon that would become GE Digital.

A few years after that, "digital" suffused the company's vast marketing operation, the adjective having swollen to crowd out all others in everything from Immelt's public speeches to the PR work of his top lieutenant, Beth Comstock. The communications office revised the boilerplate attached to every press release, the corporation's description of itself, to pronounce GE the world's first "digital-industrial" company. Immelt proclaimed in 2014 that GE would be a "top 10 software company" by 2020, with $4 billion in annual revenue from its new Predix software alone.

Spokesmen pitched stories about Immelt's burgeoning appreciation for respected consultants and writers from Silicon Valley, like Eric Ries. In a move that owed much to Welch's example, GE adapted the tech industry principles in Ries's *The Lean Startup* to its industrial business with FastWorks, an internal program that would enable a conglomerate full of painstaking engineers to replicate the failure-friendly iterative research approach of major software companies.

But most important: Immelt green-lighted San Ramon. GE wasn't cool in Silicon Valley. It wasn't of the Valley. Its claim that it was hiring recruits away from Facebook and Oracle was a fine goal, but it was still an aspirational one, no matter what the company told investors and the press. The campus it was building, to house what would be called GE Digital, was in a dead end complex of office buildings located deep in a traffic-choked corridor thirty-five miles from downtown San Francisco called Bishop Ranch. The buildings themselves, though remodeled with exposed ducts and communal kitchens, were old offices masquerading as new ones. The office buildings had once housed the headquarters of

the phone company Pacific Bell. According to local rumor, the complex had inspired the setting of the comic strip *Dilbert*.

Meanwhile, some of the tech innovations GE was producing were less real than they seemed. Back at the corporate headquarters in Fairfield, a High Impact Innovation Team (HIIT) was tasked with developing apps for use within the company — such as an iPad app to connect lighting engineers with the lighting sales force, or a web-based application for performance reviews. The innovations were dutifully pitched for small features in the business press by the division's public relations staff. These pitches were, one employee said, "very friendly and bubbly and full of shit."

In early 2014, the HIIT team received an urgent call: it was time to present one of the apps to Immelt, whose interest in tech was growing along with his public invocations of the "industrial internet" that GE would build for the world. But there was a small problem: the app in question wasn't built yet. The team had digital design files — mock-ups of the way the program would eventually look — but nothing running on a real machine to show the CEO.

Supervisors waved away these concerns. It was time, they said, to "fake and bake." Designers in Fairfield went to work animating flat visual designs, inserting enough movement to give the appearance, for one PowerPoint presentation, of an application that was working.

Word came back the next day. Immelt had loved it. (The app, in the end, went nowhere.)

In the meetings of the GE board of directors, all these machinations were regularly awarded a blessing of silence. The board, made up of current and retired business executives and academics, liked Immelt and didn't want to challenge him. They were taken, as a group, by his gargantuan work ethic, his optimistic spirit, and his eagerness to articulate far-reaching visions about where the company's strategy was taking them all.

It was no different with software, about which Immelt spoke with the zeal of the converted.

"What he says about software, in the boardroom anyway, is that it's directionally the right place for GE to go," a former director said.

"If GE is going to get where it needs to get," Immelt would say, "then

we've got to be serious about it. We got to go for it, guys. This isn't a five-year spend; it's 'get in place as quickly as we can.'"

The GE board liked this kind of urgency, but didn't usually force Immelt to commit to specifics. The board tacitly blessed Jeff Immelt's digital dreams and his plans to stand up a software company inside General Electric, but never decided — much less voted on — a critical question about such a massive experiment: how much money were they going to spend on this?

32

DINNER IN PARIS

THE EFFORT TO execute Immelt's pivot was still gathering force, which meant renewing the search for new sources of industrial revenue. The notion of buying Alstom, the French power equipment rival, had caught on quickly in Fairfield. But the rationale of the transaction, though it hit some thematic high notes, had been transformed and twisted as it wound its way upward through the layers of bureaucracy that weighed any major GE decision.

Just as GE had promised to redouble its efforts industrially, the deal represented a way for the company to invest in its oldest and largest businesses, the ones with the longest lead times and theoretically most stable returns. Alstom and GE were already something like second cousins, thanks to their shared historical connections to Thomson-Houston, the electric company that merged with Edison General Electric in 1892 to become General Electric. And to Immelt, a singular appeal of the deal was that it would give GE a dominant position compared to the other major players in the gas turbine market, especially Siemens and Mitsubishi.

When Immelt explained any given strategy or projected the company's future performance, he continued to lean on athletic ideas of conquest. His competitive instincts in business drove him to such thinking. Like a football coach intent on picking up yards on the field, he liked taking market share, and one of his favorite tropes was promising that

GE was poised to "win." Buying Alstom would drive the other teams backward as GE claimed more and more of the field.

Furthermore, Alstom now seemed eager to make a deal. As GE teams mulled the possibility of an acquisition of the French conglomerate's power units, Alstom's imperious chief executive, Patrick Kron, reached out to Immelt. Would the American be willing to meet over dinner in Paris?

Whether Immelt fully recognized it or not, Kron was increasingly desperate to find a buyer for his company. Beneath the surface at Alstom, rot ran dangerously deep. Its power business was far behind GE and Siemens in bringing a new family of turbines to market. Worse still was the unprofitability of what it was already selling. The quiet capital raise that Adam Smith had noticed had turned out to be exactly the distress signal he had suspected. Alstom was dangerously low on cash.

Now its sell-side market analysts were getting suspicious. Out in the power markets, the company was aggressively bidding for any new business it could win, often offering prices to build new coal- and gas-fired power plants so low that it would prove impossible to turn a profit once all the work was complete. But those long-term profits were of little concern to Alstom's sales teams — they needed to win those deals for the down payments alone: the deposit paid by a customer on signing such a contract was virtually the only cash coming in the door.

Kron had already wooed his intra-European rival, Siemens's CEO Joe Kaeser, at a private dinner, but had not come away with a deal. Teetering on the brink of insolvency, the French company would probably need to seek shelter in bankruptcy or a government rescue if something didn't happen quickly.

Complicating matters further was Kron's biggest shareholder: the family-controlled Bouygues SA, a publicly traded conglomerate with sprawling interests of its own in businesses ranging from heavy construction to telecommunications. Since 2006, Bouygues had owned a nearly 30 percent stake in Alstom, which it was now eyeing nervously as the company's finances looked more and more alarming. Bouygues itself needed to find a way to refinance its own struggling telecom unit. Any deal for Alstom would have to come at a price that could placate Bouygues. Moreover, there was an even more critical party to be ap-

peased: the French government, which viewed Alstom much as GE was viewed in the United States — as an industrial franchise central to the nation's identity, as well as a huge employer.

All of these obstacles were in the air as the GE corporate jet descended into Charles de Gaulle Airport in February 2014. Jeff Immelt was taking a small detour on his way to the Winter Olympics in Sochi.

He had dinner plans in Paris with Patrick Kron.

Alstom wasn't Immelt's only target. As always, multiple business development teams from the industrial units were jumping at the chance to bag the big deal that the CEO wanted. After leaving Paris on the GE corporate jet, Immelt stopped again, this time in Helsinki, to visit Wartsila, a Finnish company that made marine engines, power generation units, and oil- and gas-extraction equipment. Immelt and the GE board were also weighing a deal code-named Project Lion: the purchase of an oil and gas company roughly the same size as Alstom, according to people who knew about the deliberations.

By the early spring of 2014, the momentum began to build behind the Alstom deal. That was when, on an early conference call with some of the highest-ranking executives from corporate and the Power business, Adam Smith sensed a problem. Listening in on the discussion about GE's attempts to come up with a valuation of Alstom, Smith paid close attention when the conversation shifted to the enterprise value they would ultimately have to assign to Alstom and its assets. "Enterprise value" is a financial measure of a company's market valuation plus debt minus cash. It was a more complete valuation than simply totaling the value of outstanding shares, because it considered the entire capital structure.

Because GE wanted to take over Alstom's gas and wind turbine and electrical grid businesses, leaving its passenger train operation behind as a stand-alone French company, the American firm wouldn't be making a bid of cash per share to Alstom's board. Instead, GE would be coming to the table to negotiate an enterprise value of what it would be taking away from the existing Alstom — the value of its backlog of power plant orders, its plants and physical assets, its debts and liabilities, and its technology.

Even from the early days, according to some members of the deals

team, they were evaluating Alstom with unusually little insight into the inner workings of the French company. A simple scan of headlines would show that Alstom's business prospects had been dim of late, and that it faced questions on several fronts about its business practices, including a foreign bribery investigation by the US Department of Justice. And GE's access to Alstom's internal records, which would have shed light on the value of its backlog and the extent of its exposure to criminal liability, had been limited.

Now, as the GE executives conferred, the conversation was unfolding roughly backward. They wondered about the price that GE would need to put on Alstom — the deal's enterprise value — to put the company into play. And in particular, they tried to figure out what they would have to offer to get Bouygues enough of a return on its equity in Alstom that the majority shareholder would be willing to consider a sale.

In that moment, Adam Smith was listening to the logic of the deal collapse. Alstom had been worth examining because it was financially imperiled; the value of pursuing it in the first place had been to capture its assets at a bargain, strip what GE wanted, and then sell the rest, like an old car, for the parts. Now the company was talking about making sure Bouygues saw a return on its investment in a company that, absent a buyer, seemed to be headed for bankruptcy or a possible government bailout. Going into the deal with that kind of logic, Smith thought, GE would be seriously overpaying on almost any price it offered.

Years later, top GE executives, including Immelt, would say that they never heard any serious dissent about the Alstom deal. If anyone had harbored second thoughts, they would say, those thoughts were arrived at in hindsight. Anyone who thought the company wasn't doing what it should, or was paying too much, had been duty-bound to say something at the time rather than keep silent.

But in fact, some did attempt to raise objections, though they did so advisedly, cautiously, and briefly. A member of the Power business team that worked on the deal said that by then it had become clear that when any acquisition was a top priority of corporate, or of Immelt's, any dissent was unlikely to alter GE's course. Indeed, such a move was certain to brand the dissenter as someone uninterested in helping the team win.

With its backward logic, the deals team was working to put together a value for the entire Alstom power enterprise from which they could

back out a price per share that would get Bouygues to the table. The bankers scribbled their calculations, using values that fluctuated from €30 a share to as high as €34, as they tried to make sure that the accounting of the enterprise value they finally alighted on truly made sense — or at least could be framed as reasonable when presented to shareholders.

That effort led to some strange arithmetic. In the final GE bid for Alstom, more than half of the enterprise value — the price GE would actually pay for the struggling French assets — came from "cost synergies." In other words, more than half the value of what GE was buying would come from slashing employees and factories and other costs once the deal was closed. Cost synergies were a common part of any major acquisition, though some GE competitors, like Honeywell, made a point of not allowing their deal-makers to use the promise of future cost-cutting savings to make the math on a deal work out. In contrast, GE's entire Alstom deal, even on paper, relied on shedding workers and factories, thus raising profit margins, in order to remain in the black.

It all came back to the way the negotiation had started. "The valuation was artificial," a former GE executive said. "It was fixed based on Bouygues." Once the company leaders, from Immelt and Steve Bolze on down, had decided that they needed to make the French investor whole in order to get the deal done, the secret work being done by bankers and lawyers was an exercise not in calculation but in justification: they were coming up with an explanation for the amount the company had decided they needed to pay.

ONE DAY IN CHICAGO

THE PROTESTERS' SIGNS carried messages like PENSION FAIRNESS and RAGE as they converged on the entryway of the Sheraton Hotel and Towers on Chicago's Water Street on April 23, 2014. In their ranks were GE workers and retirees, former employees of the Erie locomotive works and the turbine halls of Schenectady, and veterans of the engine factory in Lynn, a site that predated the company itself.

Their rage was focused on Jeff Immelt, Jeff Bornstein, and the board of directors, the authors of the corporate cost-cutting program that the GE executives called "Simplification." That program had zeroed in on worker pensions and retiree health insurance as a good place to tighten the company belt. The company said that cutting unsustainable costs and trimming expenses would boost profit margins and placate investors, who remained unenthusiastic about GE's performance in the sixth year since its near-death experience in the financial crisis.

But while GE was making moves to change the terms of the lifetime health coverage it had promised workers, Immelt was also showing signs that he was in a spending mood. GE had been in the market for "bolt-on" acquisitions for the past few years. It was also keeping to its target, the two Jeffs had told investors as recently as that spring, of raising $1 billion from selling off businesses for which it no longer had any use. But Immelt had also let slip — in a classic example of GE's investor

semaphore — that the company would be willing to exceed its acquisition targets and make a much larger deal should the right opportunity present itself.

Inside the Sheraton, Immelt did his customary grip-and-grin circuit through the crowd in the moments before the annual shareholder meeting began. Towering over nearly everyone in the crowd, the CEO grinned broadly, his eyes wide in his broad, fleshy face, and squinted slightly as he greeted share owners. Immelt was fundamentally genial and greeted even adversaries like Dennis Rocheleau — a former GE labor negotiator and a current advocate for retiree benefits — with a chuckle. It was the same Dennis Rocheleau who had almost come to blows with Immelt while rowing off Hurricane Island years before. Nonetheless, as he sped through the room with a smile that sometimes seemed on the verge of a grimace, he greeted everyone he could, including reporters and some of the dissidents who would condemn the company's course during the public comment period that was soon to begin.

The shareholder meeting was a chance for the broad GE constituency — Immelt's voting public, for lack of a better term — to air out its every last disappointment with the company over the almost fifteen years of his tenure. The annual meeting was the worst day of the year for Immelt, who was powerless against whatever dissent was expressed. He had to listen, but his impatience betrayed him when he pointedly cut off speakers who belabored their point with an abrupt "Okay, thank you."

At this year's meeting, beyond the ceremonial duties Immelt would perform on the dais, fielding questions alongside Bornstein and general counsel Brackett Denniston, Immelt had something else on his mind. The deal that would be the centerpiece of his pivot was finally coming together in secret meetings in hotel suites mere blocks from this conference hall, where retirees deigned to give him hell about wanting GE to return to its industrial roots and treat its workers as they had in the good old days.

Ironically, that was exactly where he was trying to go, not that anyone wanted to give him any credit for it. Very few in the room knew what Jeff Immelt had been up to in Chicago that week, in the busy hours before he entered the great hall to address his shareholders.

Holed up in a hotel suite across town, GE's deals team and negotia-tors from Alstom were combing through asset lists and pricing models, trying to pencil out the arithmetic that would turn Jeff Immelt's and Pat-rick Kron's deal dream into reality. The storied Alstom train business, maker of the vehicles for France's high-speed TGV, would stay put, be-coming the new core of the public Alstom SA. GE would sell Alstom its own railroad signaling business, a relic of a time when GE's locomotive and railcar business was far more central to its operations. The bankers argued over the valuations they attributed to Alstom's assets; for exam-ple, GE's team felt that Alstom's murky backlog of power orders had not been reviewed comprehensively enough to make a proper estimate of its flaws and liabilities. GE board members, whom Immelt had alerted to the possible deal in February, believed that they were on track to com-pleting a $12 billion deal.

As the GE shareholder meeting approached, a transatlantic flight touched down in Chicago. Kron himself had arrived. Immelt and Kron retired to a private meeting in a hotel suite, their presence known to in-ternal business development teams and now to investment bankers who had been lined up to help close the largest acquisition in GE's history — one that, if successful, would help exorcise the ghost of Jack Welch's debacle thirteen years earlier when his attempt to buy Honeywell was blocked. The two CEOs negotiated as their teams waited for news. Fi-nally, they shook hands.

Jeff Immelt emerged with his prize, still a secret, having agreed to sweeten the pot by roughly $1 billion. GE and Alstom negotiators re-turned to work, penciling out the terms of a deal that would place an enterprise value of $13.5 billion on Alstom's portfolio of gas and steam turbines, windmills, and electrical grid equipment. The figure would be enough to win endorsement from Alstom's board and, Kron was sure, an approving vote from Alstom shareholders. Crucially, the deal could go forward with the blessings of almost all of the key parties because $13.5 billion was enough to get Bouygues on board.

Now, the Alstom CEO said, it was time to give the French govern-ment, especially its activist economy minister and markedly pro-busi-ness Socialist president, a friendly heads-up that one of the country's most venerable industrial companies was preparing to sell itself to the

Americans. Jeff Immelt, for his part, headed over to the Sheraton for the shareholder meeting. Sitting on the dais, humoring his interlocutors with attention in the public comment period, he had the air of a man with many important things going on elsewhere.

It was almost time to unveil his biggest maneuver yet.

34

A MODEST PROPOSAL

A T A LITTLE after 5:00 p.m. EST, that same day, the headline flashed across Bloomberg terminals around the world: General Electric Company was in talks to buy Alstom SA for more than $13 billion.

GE's board and Alstom's top executives and investors knew that the talks were happening. But Kron hadn't told Alstom's board and, more important, hadn't alerted the French government.

The headline would come as an unpleasant shock to one person in particular: a French bureaucrat named Arnaud Montebourg, the minister of the economy, industrial renewal, and digital affairs in the government of President François Hollande. The Hollande administration, and the Socialist Montebourg in particular, were incensed at the news, feeling blindsided to learn that one of France's most prominent investors and oldest industrial companies — a firm bailed out just a decade earlier by the government — had been secretly negotiating to sell to the Americans.

As the news reverberated, sending Alstom's share price surging more than 10 percent, Kron dashed back to Paris for a hastily scheduled meeting with Montebourg on April 24, at which the CEO hoped to placate the minister and win his support for the deal. Kron had searched the landscape of French businesses for a potential merger target that would help him solve Alstom's problems by eliminating its crushing debt and providing enough synergy for the company to stand a decent chance

at returning to health. He'd considered Areva, which made nuclear re-
actors and was already partnered with Alstom, which made the mas-
sive steam turbines used in Areva's plants. He'd also considered trying to
merge with the aviation company Safran, which happened to be a part-
ner with GE in one of its most successful joint ventures, a jet engine–
making operation called CFM. But none of the deals had come to frui-
tion. There was no other French institution big enough, healthy enough,
and willing to come to Alstom's aid, the CEO believed.

As Kron hunted for a suitor in the beginning of 2014, when his stock
price finally began plunging, as Adam Smith had expected, Joe Kaeser
appeared. The cocksure chief executive of Siemens, eager to strengthen
Siemens's position in the global power market, was sniffing around for a
deal himself. He also had a problem of his own — the dimming outlook
for Siemens's passenger rail business, which he thought could be stron-
ger if it was combined with Alstom's. Bad blood lingered, however, be-
tween the French and German conglomerates, including between the
executives at the top.

Kron believed that Siemens, under previous management, had tried
to break Alstom apart during its brush with insolvency in 2004, attack-
ing it on anticompetitive grounds before European Commission reg-
ulators. Now, without telling his board, Kron had taken the meeting
with Kaeser, but not the deal. Instead, in February, he had sent up a sig-
nal that he'd be open to talking to GE, which was hard at work trying
to find the transformational deal Immelt so badly wanted. The CEOs
had begun their courtship at that dinner in Paris, under the noses of
Montebourg and Hollande, who were trying to remake the image of the
country's Socialist Party into one more open to big business and mul-
tinational companies. But neither Montebourg nor Hollande was eager
to see an American company come in and begin laying off thousands of
French workers to boost its own profit margins.

Now GE's top in-country executive, Clara Gaymard, began a quiet
but intense lobbying campaign to reassure officials in the Hollande gov-
ernment that the American company didn't intend to gut Alstom. With
all trading on Alstom halted, the outlines of the deal became clearer
— and the fight coming up even more so. Immelt and his top advisers
boarded the GE jet for Paris, bound for a Sunday meeting with Mon-
tebourg and a sit-down the following Monday with Hollande himself.

Along for the ride were Steve Bolze, the CEO of the Power unit, and John Flannery, the company's top deal-maker.

Even as they were preparing to explain a profitable rationale for the deal to their own investors, the GE executives needed to convince the French government that the deal was the best outcome for Alstom, its workers, and the country. They also needed to secure France's help in getting the deal past the hurdle that had blocked GE in the past: the competition regulators of the European Commission. For Jeff Immelt, it was a sales call, one in which he would dangle the prospects of a strengthened power equipment maker that would be supercharged by its absorption into one of the world's most famous companies and go on to defeat competitors across Europe and Asia and even add new jobs.

They didn't make the sale.

Even before Immelt reached the Élysée Palace, Montebourg struck, pronouncing the GE deal "unacceptable." Even more worrisome for the Americans, Montebourg formally invited Kaeser and Siemens to make a bid of their own for Alstom. The German CEO was added to Hollande's schedule for his own face-to-face meeting with the president at the palace.

GE and Alstom were nonetheless forging ahead. On Tuesday, April 29, General Electric's board of directors convened via conference call. After a typically bullish presentation by Immelt about the deal and all it could mean for the company, the board voted. As usual, the decision was unanimous. GE was going to buy Alstom's power business.

"This is not our first deal in France or our first deal in Europe," Immelt told investors in a conference call on April 30, the day Alstom's board voted to endorse the deal. The company was certain that it could wring out the cost savings and layoffs it needed from Alstom's bloated structure, and that it could quickly win the necessary approvals from regulators in Paris, Brussels, and Washington.

"Over time, I want industrial to be bigger, and I would say this gives us an opportunity to reset the bar at 75–25," Immelt said, meaning that industrial units would make up 75 percent of GE's earnings and just 25 percent would come from finance. "I think this gives us a chance to reset the bar and do it in a very profitable way."

But they were underestimating Montebourg's wrath. As April turned into May, the minister pushed GE to delay the final shareholder vote, in

which Alstom investors would have to sign off on the transaction, and extracted a pledge from GE to add one thousand jobs to the power business in France.

GE executives and the bankers weren't too worried about avoiding job cuts in France, since so many of Alstom's workers were employed in other countries. Power CEO Steve Bolze wasn't concerned either. It wasn't his problem, really. As he told people, GE could simply hire hordes of software engineers to work in Paris to support Predix. For the head of a division trying to close a massive deal, Bolze didn't have to worry about such details because GE was big enough to absorb that cost without flinching. The cost of meeting the key commitment in the deal wasn't even going to fall on the Power division.

Montebourg attacked Kron as well, accusing him of a "breach of national ethics" for pursuing the Alstom sale in secret, and in an appearance in Parliament, the minister asked whether the CEO would need to be connected to a lie detector. On May 5, Montebourg wrote to Immelt, on Hollande's behalf, threatening to block the deal. Less than two weeks later, Montebourg's hand grew stronger. The minister had finagled from the government a *décret Alstom*, which broadened the government's right to block foreign acquisitions of French companies in critical sectors. The categories in which the minister could act were expanding from national security and nuclear interests to makers of other electric generation equipment as well as the healthcare, telecommunications, and water industries. And Montebourg continued to welcome the ominous noises coming from the direction of Munich. Kaeser said that Siemens was strongly considering making its own bid for Alstom's power assets, and Montebourg urged the Alstom leadership to wait until they could consider the competing offer.

The world could see what Immelt hoped to do: lock the competition for decades into a second tier by gobbling up all the market share in power generation. Siemens and other competitors, like Mitsubishi, weren't about to let him get it without a fight.

And Montebourg, a hardened skeptic of American corporate giants like GE, feared that one of the most important national institutions was rushing itself into a fire sale to a cold-hearted company best known for its facility with layoffs and transactional churning.

GE would not simply waltz away with Alstom on Montebourg's watch.

35

"WE KNOW WHAT WE'RE DOING"

JEFF IMMELT RESPONDED to the French fuming with his customary chuckle. "Trust us," he told a group of financial analysts in late May, referring to the company's long experience of navigating the French bureaucracy thanks to the facilities it maintained there. "We know what we're doing."

But the French demands were serious, and they required a change in GE's negotiating posture. That week GE said that it would extend the deadline for Alstom to approve its offer until June 23; that gave Siemens enough time to put together a bid if it was serious about foiling Immelt's plans. The GE press office whirred at an ever-higher frequency, spinning the paradoxical narrative that the opposition from major political figures was run-of-the-mill deal stuff, hardly newsworthy, while also trying, sotto voce, to impress upon reporters how truly great an opportunity GE's offer presented for the French company and the French economy.

GE's image stewards didn't like Immelt being described as a salesman, though he was, but every one of them believed in his ability to lift the mood in a room and to sell even the most recalcitrant customer when he turned on the charm. The boss, his big deal hanging in the balance, would try to turn the bright lights on the French public itself to swing support GE's way. On May 27, a day ahead of another planned meeting with Hollande, Immelt went to Paris to pitch Parliament on what he now called an "alliance" between Alstom and GE's power units. His tes-

timony had a note of defiance, reflecting how deeply frustrated Immelt was with Montebourg, Hollande, and the Parisian bureaucracy. Nevertheless, he was full of reassurance that tying up with GE would be good for the French company, not just for the American one. "We come to build," he said. "We don't leave when times are tough."

Siemens's bid arrived in June, hand-delivered by Kaeser to Kron and the Alstom board. The German firm proposed buying Alstom's gas turbine business, while Mitsubishi Heavy Industries (MHI), another competitor, would take minority stakes in Alstom's steam turbine, nuclear, hydroelectric, and grid businesses and agree to partner on future research and procurement. The deal would have seen Kaeser swooping in with what he said was a higher implied value for Alstom than what GE was offering, and the infusion of €7 billion in cash would have wiped away Kron's debt problem. Even more, to the delight of French officials, the Siemens bid would have kept the company in France and guaranteed the hiring of one thousand workers — a perk that raised the pressure on GE to formally add a job creation guarantee to its own offer. Kron promised a careful review of the Siemens offer, even as GE took to the press to bad-mouth it as equivalent to a breakup of the French conglomerate.

By late June, GE's deals team had been huddling for weeks, working on a way to keep the financial outlines of the Alstom gambit roughly what they had briefed for investors but tweak the offer enough to make it seem more attractive than Siemens's bid. Immelt had spoken of an "alliance" in his remarks to the French Parliament, acknowledging that the greatest anxiety provoked by the GE deal, beyond the simple threat of job losses, was the fear that GE would be appropriating and essentially making off with the multigenerational assets of a French company. His team's solution would be to keep French ownership in parts of the company, enabling Hollande and Montebourg to say that their intervention had retained French legacy assets. But this would work for GE only if it could retain the leeway to run the business as it had intended — leaner, and probably with a much smaller footprint and workforce than what Alstom had limped along with for so many years.

Immelt was ready to go public with GE's revised deal. They would form a triad of joint ventures, not unlike the structure of the Siemens-

MHI proposal. Each joint venture would hold an Alstom electrical grid, renewable energy, or nuclear power business, with the ownership shared fifty-fifty by GE and Alstom. The French government would take a preferred share in the nuclear venture in order to guard it against outright takeover by a foreign entity. And the French government would buy up a 20 percent stake in the new Alstom, mopping up most of Bouygues's ownership stake in the company and giving the government new powers to dictate the French company's future strategy.

Siemens too could see that it was now GE's deal to lose. To some extent, Montebourg had prevailed. GE had altered its terms, sweetening the pot — though not necessarily for Alstom and its shareholders, who were desperate for any buyer or merger to prevent what was a rapidly approaching financial collapse — but for France. Montebourg had asserted a leadership role for the government in the transactions of companies (those of one more than the other) that thought of themselves as larger than and unconstrained by the state. Montebourg's stalling of the closure of the deal, his decision to invite Siemens's bid, and his interventionist posture had all compelled GE to alter its terms. The company of Boulwarism ("our first offer is final") had blinked.

But now Montebourg and Hollande bent to other European realities. There could be no further playing out of the string, no counteroffers from Siemens. Montebourg would say in later interviews that the European Union's antitrust rules, he had realized, were too strict to permit a tie-up of a major French firm and a major German firm. He had hoped to create European "champion" firms to compete globally in electricity production and transportation, but the EU's inward-focused anticompetition schema had foiled that plan and forced France to allow GE to take Alstom over, Montebourg told the *Wall Street Journal.* "All of the operators are ill," he said, in a bleak assessment of European corporations, "and the consolidation is impossible."

That left GE, and this was where Immelt won. Hollande and Montebourg did need a buyer. Alstom was teetering, far more than anyone was letting on. And so the French government, having extracted job protections and joint venture structures from GE in exchange for promises not to block the Alstom acquisition, threw its weight behind the deal to get it through the anticompetition regulators in Brussels. For the very

reasons they had been so nervous about GE buying Alstom — the social and economic effects of it being raided, downsized, or allowed to simply collapse — the French officials now needed the EU to let the deal go through.

On June 20, 2014, GE won the government's assent, and Alstom's board voted in favor of the acquisition. Pending approval from the European Union, Jeff's big deal was going to happen.

36

BUY HIGH

FOR ALL THE leaders' enthusiasm about the digital realm, GE continued the endless job of refreshing and reshaping its heavy industrial business units. While Immelt dreamed of creating a heavy-industrial Google in San Ramon, he also gazed longingly on the vast oil industry conglomerates like Schlumberger and Halliburton, whose fortunes from both equipment and oil-field services had soared during the long bull market that began after the financial crisis and ran into the early 2010s. Even before the crisis, Immelt and his lieutenants had begun to acquire. The company bought Vetco Gray, which made drilling equipment, for $1.9 billion in 2007, adding it to Nuovo Pignone, the Italian oil equipment maker that Welch had purchased in the 1990s. After the financial crisis, Immelt had altered GE's portfolio through subtraction more than anything else, shedding businesses that had lost their appeal after the Welch years. But now GE became, in a series of high-dollar acquisitions, a player in the oil and gas equipment market virtually overnight. They bought an oil-field service equipment business based in Britain, manufacturers of pumps, wellheads, regulators, and piping in Texas, and a Norwegian firm whose sensors measured the pressure and flow of oil.

GE in acquisition mode twirled like a tornado, gathering up smaller companies to blend into the overall mass — sometimes at prices that Immelt's critics, like sell-side Wall Street analysts and business journalists, would occasionally, quietly, criticize. Churning on, the company hopped across the landscape of suppliers to the fracking and offshore

drilling crowd — drill-bit makers, compressor vendors, pipeline inspectors — scooping up more companies to add to the mix. GE was going big on the oil business.

And then one day GE touched down in Lufkin, Texas.

The announcement came on April 8, 2013, with a GE press release that invoked the "world-class people, equipment and services" of Lufkin Industries, which GE was swooping in to buy for $3.3 billion. The Lufkin company, which made "artificial lift" equipment to help pump oil and gas up from underground reservoirs, had been in business, along with its forge, for 111 years, dominating the little city's economy. Lufkin Industries had been Lufkin's biggest employer for decades, its fortunes moving in sync with the community as the oil-producing economy of Texas and Oklahoma crested and fell over time.

Recent years had been kind. The growth of the North American shale market had driven a surge in equipment orders and driven up Lufkin's profits. (That was one of the major reasons why the small firm, when GE came calling, had been able to extract what critics felt was a steep price.)

The Lufkin deal was a late-arriving piece of the $14 billion spree that built up GE Oil & Gas, from a one-off Welch deal into one of the most important industrial business lines in the company. Although GE had owned an Italy-based oil equipment business since the 1990s, Immelt was now making oil the industrial sector into which he extended GE in such a large way. The strategy sounded a dissonant note when set next to some of Immelt's rhetoric about the environment, especially the marketing tag of "ecomagination," which GE affixed to any of its products that could boast of increased efficiency. Still, Immelt and the youthful CEO of the oil unit, Lorenzo Simonelli, were confident that they were positioning the company to take advantage of the one great need of almost every economy in the world as economic growth sped up and populations grew. GE had found a way to siphon profits from oil.

By 2014, the company had already assembled a conglomerate within a conglomerate through a spate of energy industry deals. The unit now known as GE Oil & Gas was its third-largest by sales. Oil and gas accounted directly or indirectly for one-fourth of GE's $100 billion in industrial revenue that year. While Immelt heard, and was annoyed by, the chirping of some analysts who felt that he'd paid a premium to leap into

the oil and gas industry several years after his competitors, the company's leadership was sure that the ensuing years would show the bet paying off.

Crowded into a conference room at the St. Regis Hotel in Manhattan in September of that year, GE Oil & Gas executives briefed investors and analysts on their plans for the future. A reviving global economy would push up oil demand, they predicted, compelling GE's customers to order more and more of their drilling supplies. One after the other, as the analysts sipped coffee and munched on pastries, the executives clicked through their slide decks and outlined a steady progression to increasingly ambitious technologies, all of which would win the company high profit margins and steady sales growth.

GE executives foresaw demand for oil and gas becoming so strong that it would stimulate new drilling in the most technically foreboding environments, like deepwater pockets of oil located far offshore. GE would wring out profit all along the production line, they assured the gathering, and could introduce new machines that weren't yet in use. One slide showed a massive separator, sitting on the seafloor, where it would remove natural gas from oil as it flowed out of subsea deposits, feeding both commodities to ships on the surface. There would be great opportunity for GE, executives said, as high prices drove more exploration off the coasts of Norway and Scotland.

The seafloor separator was the kind of high-tech device, intended for use in a high-risk environment, that only stood a chance of being used at all if the oil company was going to make a lot of money selling the oil it produced. In spite of the fact that oil companies, to maximize their profit, stuck to simple and cheap production techniques when the price was low, GE had a reassurance to offer. It was an executive named Andrew Way, a specialist in offshore equipment, whose job it was to present that slide. Not to worry, Way said. GE's "base case" assumption for all of the rosy pictures it was painting about its oil unit was $100 for a barrel of oil.

Brent crude had closed out the previous month at more than $105 a barrel, only a little off its summer peak. But even as GE managers were speaking, a long slide had begun. By the end of October, oil had fallen below $100 a barrel. By the end of the year, it was under $80. A year later,

a barrel of Brent crude cost less than $50 — less than half the price at which GE had just assured its investors that all its plans for the oil business made financial sense.

Immelt, Bornstein, and other GE executives soon took to insisting that they paid little attention to the price of a barrel of oil; this wasn't the market they'd been trying to elbow their way into, they said, as they built up the oil and gas unit. The explanation didn't make sense given the industry's direct link to oil prices. In reality, the company had chased a fad, and the fad had suddenly gotten very expensive. That could lead to only one thing: the beginning of the "right-sizing" of GE Oil & Gas.

As with all of GE's industrial business lines, there was supposed to be security in the vast backlog of orders taken over the years by the oil and gas unit. But all those orders for pumps and compressors and drill bits weren't going to do GE much good if the customers, pinched by the oil price collapse, couldn't actually pay for the machinery they'd bought. GE executives privately acknowledged the sensitive negotiations that were unfolding as customers sought to renegotiate the pricing on their orders, delay deliveries, or, in some cases, cancel their orders outright. GE managers in each case had to weigh the pain of taking a financial haircut on some of the $4 billion of orders in the backlog against the stability that the orders offered once the oil price recovered — whenever that was going to be.

Pressed on the need to renegotiate parts of the order book, Bornstein tended to pivot: GE was also engaged, he said, in "standardization." Customization had come at a cost to GE, and in the current environment it wasn't worthwhile just to land a given sale. There would be no more bespoke offerings for different oil producers, with equipment arrays customized to their preferences like lapels on a tailored suit. To "standardize" the operation, GE got to work cutting, starting with the area that could save GE the most money in the long run: people.

That meant it was only a matter of time before the ax fell on Lufkin.

Down in Texas, there had been worrisome signs already. Local officials had been hopeful that GE would follow through on something referred to in the press as Project Phoenix. GE was considering knocking down a chunk of the more than fifty-thousand-square-foot foundry complex, extending it with new structure, and renovating the remaining building. The plan had suggested a commitment to the business that

was still central to the town. But Lufkin leaders knew the ways of the oil patch. They knew what happened to big plans when the price of crude fell.

GE had been warning of job cuts across its oil and gas units for months, and it had laid off 575 workers at another former Lufkin facility nearby. Town officials held out hope now that GE would find a buyer for the foundry. But on August 24, 2015, GE announced a different decision. The foundry at the center of Lufkin, Texas, would be shut down.

"They haven't brought good news to our office in a long, long time," Lufkin mayor Bob Brown told a reporter that day. "When they say they're going to dismantle the foundry, that pretty well seals the deal," he added. "I just passed the back gate and saw guys walking out of there with their heads down, and that just breaks my heart."

There was little he could do, Brown acknowledged to another reporter, about the price of oil, which was the key factor that drove the decision to shutter the foundry and lay off another 262 workers across the Lufkin facilities in Angelina County. Still, the smaller company had been by some measures a victim of its own success. Afloat on fracking profits during an oil boom, Lufkin had caught GE's eye and been swallowed up at an expensive price, only to become a casualty when the conglomerate couldn't abide the hit to earnings that a prolonged dip in the price of oil represented.

Back at GE headquarters, the company was tiring of its oil adventure — or at any rate, of having to manage the long downturn without help. GE was running out of places to cut costs in the oil and gas unit to preserve what it could of its profit margins. The cuts were getting close to the bone, and soon enough, the only remaining option would be to report losses on the business. There was only so much fat to be cut.

The other major oil industry suppliers were similarly weary, trying to shore up earnings by slashing jobs, trimming project costs, and squeezing their own customers and suppliers wherever they could. (The wildcatters had it worse: many of the mom-and-pop operators of the American oil patch started to file for bankruptcy.) One year later, GE would merge its oil and gas unit into the oil-field giant Baker Hughes, keeping for itself a more than 50 percent stake in the company and spinning out a new public company to be run by Simonelli, under GE's control. The transaction eased GE's exposure to the ongoing oil rout and gave the

new company, dubbed "Baker Hughes, a GE company," vast new areas of redundant employees and operations to eliminate.

With Baker Hughes, GE changed its tone a bit. The deal was transformational, but in which intended direction wasn't made clear. GE execs like Bornstein would proclaim that the deal gave them "optionality," but the reality was that investors were left in the dark on the strategy: Was GE doubling down on oil? Or was it preparing to exit the industry? The idea of holding such a long-term option was nice, but the game pieces in the positioning were people, and those who didn't leave their job had no idea where the future of the company might be.

The new arrangement didn't spare Lufkin. The historic foundry was closed. The city's annual financial report now just shows a blank line when listing the company's employment tally, evidence of the more than four thousand jobs that evaporated after GE came to town. Between two Mondays — the day GE announced it was coming to Lufkin and the day the company said it would move on, leaving a shuttered foundry at the center of town — just 868 days had passed.

THE PROBLEM WITH OWEN'S HAMMER

MEANWHILE, IN CALIFORNIA, the pledge to turn General Electric into a major software company was turning out to be slightly more challenging than Immelt had made it sound. The corporate vision of connecting the world's heavy machinery, harvesting data, and streamlining the use of energy relied on a software ecosystem that didn't really exist yet — operating systems and apps, data protocols and standards, trouble-shooting tools for the inevitable early hiccups, and clouds and server farms to hold the massive volume of data that GE hoped to analyze.

Some executives at the company thought the obvious answer was to form a partnership. There were already massive, global companies that for decades had been building a software infrastructure to host and influence virtually every aspect of human life, communication, and business. Google. Oracle. Microsoft. Amazon. GE's knowledge of the machines it made and the needs of its customers was unmatched, but that didn't make it the most obvious candidate to construct the scaffolding of code on which the "industrial internet" would rest.

Nonsense, Jeff Immelt said. The company's mantra was "domain knowledge." It knew what IBM never would about the data that mattered from sensors in the hot section of an aircraft engine. GE knew what data it was looking for, and it would simply build the software it needed to gather, house, analyze, and use that data. In doing so, GE

would also own its platform outright — as well as this bold new frontier of the internet — without being obliged to share the spoils.

That is, once they could make the software work.

GE had first launched Predix a couple of years earlier, calling it a "first-of-its-kind industrial strength platform that provides a standard and secure way to connect machines, industrial big data and people."

It was perhaps telling that as GE rolled out a product it claimed would change the way industrial machinery and major economic sectors operated, the company's executives couldn't even agree on how to pronounce it. Immelt extolled the potential of a product he pronounced as "PREE-dix," sometimes then turning over the microphone directly to Ruh, whose version sounded more like the verb "predicts." If those discrepancies suggested a deeper miscommunication about what GE's crucial new software platform was for and what it could do, no one was raising the issue publicly yet. For the most part, Predix, while well advertised in GE press releases and investor presentations, wasn't really on the market at all.

The Late Show with Stephen Colbert debuted on CBS on September 8, 2015. The former correspondent from Comedy Central's *The Daily Show*, who had emerged as a star in his own right on the spin-off show *The Colbert Report*, emerged from a bearded hiatus regroomed, out of his parodic conservative character, and ready to lead an earnestly jokey variety show tuned for the twilight of the Barack Obama years. George Clooney showed up for the debut. The new host thanked his predecessor, David Letterman, and sang along with Mavis Staples. And at one of the commercial breaks, there, for the first time, was Owen.

"Owen" was, to a certain brand of GE executive, retiree, or investor, the end-stage product of the age of Immelt and Comstock. In a series of broadcast television spots produced by GE's longtime ad firm BBDO, the company introduced a young computer programmer, the titular Owen, who informed increasingly perplexed groups of his friends and family that he had taken a coding job — at General Electric.

The various confused responses Owen elicited in the ads (like the character's father presuming he has taken the coding job because he cannot lift a heavy hammer, or friends expressing sympathy that he hasn't been able to secure a more meaningful job, like working for an app that allows users to attach fruit to images of animals) were delivered

with a self-deprecating humor for which GE had never been known. In a media blitz aimed at rebranding GE as a "digital industrial company" in the minds of the general public — and also reminding potential job applicants that GE still existed at all — BBDO, Comstock, and her marketing team first had to confront the utter uncoolness of the parent company. Everything usually venerated by GE's corporate communications, like its carefully parsed connection to the workbench of Thomas Edison and the ubiquity and raw power of its massive industrial machines, seemed the very opposite of the tech industry in which it now wished to stake a claim.

The Owen ads spoke to some of the truth of GE Digital. Notwithstanding some claims that GE was making to investors about its success in luring young programmers away from jobs at Facebook or Apple to work in San Ramon, in fact, recruiters said, the response in Silicon Valley had been not unlike Owen's fictional brunch table. To the characters in the ads, it was much more enticing to a young programmer to work at something like the fictional "Zazzies," putting cassava melons on the heads of kittens — or the real-world equivalent thereof — than to enter GE Digital, where the challenge was not just the coding but integrating a software operation into a wary, if not skeptical, company full of engineers.

GE claimed that the Owen ads were a recruiting success, and that job applications at GE Digital rose eightfold after the series of spots were aired. "People inside the company are just in love with the campaign," Linda Boff, the chief marketing officer, told a writer for Contently, an ad firm that GE used to produce branded content. "We have brought the actor who plays Owen to some of our internal events, and really you'd think we were bringing the Beatles back together. People are so excited that here's a story about the company, but it's really a story about them. They're our Owens."

But others in the company, to say nothing of former executives and board members, recoiled. What seemed to members of Boff's and Comstock's marketing teams to put GE in a self-deprecating but still positive light seemed disrespectful and counterproductive to other executives, who didn't feel that GE needed to be marketed with an ironic wink, not even to twenty-something job-seekers with programming degrees.

There was also a lingering skepticism about the promises these ads

were selling. GE Digital spending, blessed in the abstract by the company board but left largely in the hands of Ruh and disparate digital teams embedded in the industrial units, was now soaring. GE Digital was on track to spend $5 billion by 2016 — a massive sum even by GE standards, equivalent to about half the R&D budget for a new, clean-sheet jet engine.

There was a lot of talk about Predix and everything digital. There were annual investor conferences on the San Francisco waterfront and demonstrations by major GE industrial customers, including Union Pacific and Exelon, pointing to the potential for big data in the future.

The problem was that, $5 billion in, it wasn't clear whether GE would be getting what it paid for.

As the year came to an end, Immelt planted a flag that would define the rest of his career: he declared that GE would produce at least $2 of profit per share in 2018. It was an unusually long-term projection, and its meaning was undeniable to Immelt. Hitting the target would be the proof of the pivot's success. Delivering $2 per share would not only show that the company's growth hadn't missed a beat after shedding most of Capital and consuming Alstom but also assure investors that they would all be rewarded for their patience.

"I DON'T EVEN KNOW WHAT WE'RE SELLING"

BY MARCH 2016, the marketing was in full swing: the future of GE businesses was "digital." Offering software "solutions" rather than old-fashioned service contracts was proving to have commercial appeal for investors and analysts alike.

It was March 11, a Friday, and crowds of eager executives were gathering in Lower Manhattan, queuing for elevators to reach the upper-floor meeting rooms of the New York Stock Exchange. The event was an "investor day" for GE Healthcare, the long-troubled division of the company that made X-ray and MRI machines and handheld sonogram scanners, and included, after one of Jeff Immelt's most expensive deals, a life sciences division whose future lay in diagnostics and the rise of bio-pharma. Under a new chief executive, John Flannery, the former head of business development, GE Healthcare was making its first presentation to investors in years. If Flannery could turn around the moribund health unit, he loomed as a potential dark-horse candidate to be CEO of the whole company one day.

But the day for a change of leadership at GE was still a long way off. The Healthcare business had long been troubled by forces similar to those that had led GE to give up on its signature appliance business. As outside entrants jumped into the health sector and the price of individual machines fell, the industry had become increasingly commoditized. GE Healthcare's customers — health systems and hospitals, doctors' offices fighting uphill reimbursement battles with insurance companies

or Medicare — were themselves squeezed financially. They wanted discounts, not new-fangled features and equipment upgrades that might have helped the Healthcare unit drive up revenues and the all-important earnings.

Other companies were getting out of the healthcare business altogether. Siemens spun off its healthcare machinery unit, Toshiba had left its healthcare business behind, and Philips too was tiring of the sector. And the previous GE Healthcare CEO, John Dineen, had eventually been forced out after failing to find a way to shock the business back to growth.

Against that backdrop, the wry, somewhat nervous-sounding Flannery took the microphone to regale the crowd with familiar-sounding boasts about GE's product line. One of the great strengths of healthcare was that it opened doors to markets for other industrial business lines. As Flannery described it, healthcare was the beachhead team in some markets: it may not have made killer margins selling handheld sonogram scanners for use in poor villages without major hospitals, but planting the GE flag among local officials was widely seen as helpful when it came time to make more lucrative deals for power equipment, oil-field supplies, or jet engines.

Some of the most bullish analysts thought that GE Healthcare was intriguing primarily as an asset to be spun out — to create a pure-play medical device business with relatively high margins and few direct competitors. Flannery, however, was making the case that the business should remain under the conglomerate's umbrella. He knew that it was a good business, and that selling it now didn't make sense before trying to get it in better shape. Flannery detailed this strategy to Immelt when he was running GE's deals team, and Immelt ultimately put Flannery in charge of GE Healthcare to carry it out.

And he was largely successful that day in Lower Manhattan, convincing some in the crowd that he was tougher on cost control than past leadership had been and that the dour market outlook for the next two or three years would make it essential to keep Healthcare within the larger company.

The GE Healthcare team was also pushing Predix that day. Executives spoke excitedly of its potential to harness data to improve medi-

cal research and the quality of hospital care. GE customers, including a hospital network in Pittsburgh, were trotted out to talk to investors and reporters about how software would maximize the potential of GE's big machines. Predix would serve as a software backbone allowing specialists in far-flung locales to collaborate on examining the scans generated by GE MRI machines and sonograms. Huge sets of scans from similar procedures would fuel the same machine-learning process that GE hoped to unleash on everything from aviation maintenance to power generation. With millions of examples to learn from, the company said, apps running on GE software, used as a first step in medical testing, could provide a powerful new tool to catch diseases and problems before they became acute.

But the real world turned out to be different from what was promised in that room that day — GE Healthcare's teams weren't able to make the new products work as promised. Even GE and Predix had to obey laws of physics that weren't susceptible to marketing. MRI machines (and their cousins) produce a staggering amount of data in a small amount of time. The files proved to be so big that the feats of real-time data analysis, file transfers, and machine learning that GE had envisioned happening seamlessly among hospitals and clinicians were too cumbersome to achieve quickly — if and when they could be performed at all. While the marketing and sales folks were getting ready to sign software contracts, there were still lingering and ominous internal questions about whether GE's goals were realistic, or even possible.

But these goals were still far off in the distance, the marketing language notwithstanding. Teams in San Ramon and in the digital outposts within GE industrial businesses were still figuring out what they could build, and in the meantime they found themselves essentially giving software programming jobs to individual GE customers. Building customized code for individual customers who were willing to give Predix and its add-ons a try was slowly providing learning opportunities for GE Digital. But the company was unlikely to turn a profit from customized code, since it usually couldn't be sold again to someone else. Software businesses usually turn a profit by selling uniform programs that can be sold many times over once they are built, not by designing unique programs, customer by customer.

But even making those bespoke sales required something else: a viable product that GE's customers — oil drillers, airlines, power plant operators, freight railroads, hospital chains — actually wanted.

GE didn't just pour money into Predix — it smothered the project with cash. But without a coherent strategy and well-thought-out processes, the product development path was a wasteful one. GE's plan to move fast, produce a viable product, and then perfect it in the field got bogged down partly because of the size of the effort. GE hired armies of new employees and gave them all the resources they wanted to build its vision. It was like an auto company building an assembly plant, hiring workers, and leaving them standing on the production line, waiting for the vehicle to be designed.

Instead of charging a small team with developing the best product and then letting the operation grow with the product's evolution, GE set up a huge organization that wasn't quite needed yet. Development was often paused or delayed in order to start the process over entirely or just to stabilize the systems.

Inside GE, there was pressure from corporate management to use the software and to show results. While some divisions were hesitant to commit to adopting a program that needed a lot of work, others were going in a different direction to develop their own software tools. Immelt pushed back on business leaders, exhorting them to get off their high horses and stop complaining. As one executive said, "There was a lot of talk of saluting the Predix flag."

Naturally the marketing and sales teams were concerned that potential customers would see that GE wasn't even using its own software. Squadrons of salespeople had been deployed to sell the vision, but what customers wanted to see was the proof behind the concept. However, there was little to show. Once again, the marketing was getting ahead of the product. In fact, the sales teams weren't even entirely confident about what their product could do. Instead of knowledgably hawking GE's lineup of products and services for a given division, they were now pitching customers on a deep analytic software platform that was hard to understand and harder to explain.

As the problems deepened, it appeared that no turning back would be allowed. In a meeting of GE senior executives, one executive ques-

tioned whether it even made sense to move forward with Predix. Immelt fumed at the suggestion of abandoning the work and quickly nixed the proposition, making it clear that the direction wasn't up for debate and the marching orders were still the same: go build it.

GE wanted the system to work with everything, so customers around the world could just connect to GE equipment and go. Engineers were concerned, however, that even though Immelt and upper management were sold on the Predix idea, management's lack of understanding of software development was holding back its growth. GE also planned to build its own data centers, from poured concrete foundations to roofs. The idea was to have a GE-owned and -operated cloud for its customers' data, but building such an operation from the ground up would be hopelessly slow to achieve and wildly expensive. Besides, companies like Amazon and Microsoft were already pouring billions into providing just such services to other businesses. Why would GE, arriving late to the business, try to duplicate their offerings?

There was also the simple problem of trying to put a lot of things on the same platform. Engineers found that the tiny sensors in GE's machines were producing tons of data, but since they used distinct coding on systems spread throughout GE's global businesses, putting everything on the same platform made the functioning of the apps excruciatingly slow.

At the same time, GE's venture capital operation was buying stakes in companies that developed various tools that could be used with Predix. The deals added new capabilities to the division, but they also made the conglomeration of Predix's code even messier. The result was software with plenty of bugs, a difficult user interface, and none of the requested features.

Finding itself unable to compete, GE would eventually shift its strategy. Meanwhile, GE had essentially put a cake in the oven and then offered it to expectant diners just five minutes later. The process produced an abundance of grand promises, deep costs, and employee confusion.

Down at the New York Stock Exchange, GE Healthcare's investor meeting had broken up, and scores of fresh-faced GE executives were crowding into the wood-paneled elevators with investors and the occasional reporter. As the doors closed on the seventh floor, a *Wall Street*

Journal reporter was backed against the rear wall as two young male Healthcare employees loudly chatted about the business's digital offerings in the front of the stuffed elevator car.

"Have you guys figured out your problem on pricing?" one asked, and the other ruefully chuckled.

"I don't even know what we're selling," he replied.

39

JEFF'S DEAL

A YEAR LATER, IN the spring of 2015, Adam Smith's team was getting nervous. They had known that GE would have to agree to shed some pieces of the Alstom conglomerate to get the major deal through regulators around the world, especially in Europe and the United States but also in Asia and South America. Still, the price that was being asked in concessions was starting to bump up against the logic of the deal GE had pitched to its investors.

European regulators, intent on preserving competition on the continent, had ruined GE's plans before. In the final days of Jack Welch's tenure in 2001, the European Union commissioner for competition, Mario Monti, had foiled Welch's attempt to complete an enormous merger with Honeywell, the aerospace conglomerate. Intended to be Welch's crowning achievement, that deal had been one he wanted badly enough that he had stayed on past his intended retirement date, forcing Jeff Immelt, the heir apparent, to stand awkwardly on the sidelines as the old boss tried to get the deal approved.

The Honeywell deal also would have achieved something similar to what Immelt was after in 2015 — a major expansion of one of GE's key industrial units that would have enabled the company to both increase its market share and boost its gross profit margins through the aggressive cost-cutting and layoffs for which it was known. But Welch had miscalculated badly the resistance he would face in the European Union, whose bureaucracy was eager to assert its authority to police an expan-

sion of an American quasi-monopoly that might threaten European competitors and customers like Airbus. The required concessions were too much, and Welch saw it was time to walk away from the deal. Monti's blocking of the Honeywell deal had been the lone sour note during Welch's well-timed exit from GE into the pantheon of executive pundits in 2001.

Now Immelt needed to avoid the same fate with his capstone industrial deal. To do that, with lobbying assistance from the French government, including the new economy minister, a dashing ex-banker named Emmanuel Macron, GE would have to negotiate concessions that wouldn't invalidate the logic and math of the deal they had labored to strike with Alstom and the leaders in Paris. That would turn out to be far harder than GE's leaders expected.

From the outset, the company faced a tough adversary in Margrethe Vestager, the European Commission minister who now held Monti's old job. European regulators were deeply concerned about what consolidation in the power market could mean for the European Union countries. While French officials had focused on French jobs and ensuring that nationally important technologies remained in the control of French companies, the continental regulators had a different mission. The elimination of Alstom as a competitor in the power turbine business would constrain every utility and power generator in Europe and the rest of the world by shrinking an already small universe of options from which to buy a new natural-gas power turbine or to have one serviced over the decades. There would be only three major competitors — GE, Siemens, and Mitsubishi-Hitachi — to choose from, along with a handful of smaller outfits, like Ansaldo of Italy and the steadily growing but not yet widely used offerings of Chinese firms like Harbin.

The deals teams huddled between meetings with regulators in Brussels, trying to calculate the value of what GE would need to give up from Alstom's assets in order to win the European Commission's assent. It was clear early on that any deal would have to include some way to preserve a European alternative to GE for customers seeking to build new power plants.

The team had a problem in the United States as well. The Justice Department had zeroed in on a Florida-based subsidiary of Alstom called PSM (Power Systems Manufacturing). This small unit wasn't one of Im-

melt's bragging points about the deal he had struck. The CEO preferred to talk about the global field of play and winning new market share across the world. But to some of the bankers working on the deal, PSM seemed like a silver bullet for GE's business. This Alstom unit sold parts and services for existing gas-fired power plants and was one of just three main providers of parts for GE's 7F turbines. But because those turbines made up 70 percent of GE's installed base of turbines, the government argued, swallowing up its primary competitor in the spare parts market could give GE overwhelming power to set prices for those critical parts, in a direct violation of antitrust law.

Even more valuable in the minds of some GE Power executives was the fact that PSM licensed patents to provide parts and services to turbines made by other manufacturers, including Siemens. That meant that the little Florida unit could serve as a pathway to the area of the power business where virtually all of the profits lay. Most power turbine companies lost money on an accounting basis from selling the machines themselves; after all, a new GE H-class turbine took months to assemble and required a huge investment that wasn't outweighed by the purchase price. But virtually no power plant operators would undertake such a massive purchase without also arranging a long-term service contract with the manufacturer to keep that turbine running. It was those contracts — multiyear, multimillion-dollar maintenance deals with a certain number of guaranteed operating hours — that turned the fleets of turbines produced by GE and its peers into profit engines.

Taking control of an entity like PSM would enable GE to sell services not just to buyers of its own turbines but to those who had bought turbines from competitors like Siemens. To the bankers working the Alstom deal, this was one of the potential blockbusters buried beneath the headline. GE had an opportunity to turn on a river of profits at the expense of every other power turbine maker active in the United States.

Besides the antitrust question, there was another wrinkle emanating from the Department of Justice: Alstom's crimes. Even as GE tried to close the books on its acquisition, a multiyear federal investigation was nearing its conclusions about the tactics Alstom had used as it tried desperately to close deals overseas and bring in cash it badly needed. The company, the Justice Department found, had engaged in rampant bribery abroad to win power deals, a violation of the Foreign Corrupt

Practices Act (FCPA). In the government's telling, Alstom's violations of FCPA had been systematic and egregious: the company had paid more than $75 million in bribes over a decade to win business in countries such as Saudi Arabia, Egypt, Indonesia, and the Bahamas.

With the deal already in the works, Alstom's guilty plea attached another string to it: the money to pay the hefty fine was to come from Alstom, not from General Electric. "That was something we insisted upon," the head of the Justice Department's criminal division said when the deal was announced. A GE spokesman told the *Wall Street Journal* that Alstom's liability had been well known and included in the purchase price. But by 2015, with Alstom's existing business withering after months in deal limbo, it was clear to everyone involved that the company wouldn't have the cash to cover a nearly $1 billion fine. So that challenge too sat in the pile of outstanding issues before the deals team. Somehow they'd have to find a way to get the Justice Department its money.

The cost of the concessions continued to mount. GE would be getting even fewer assets than it had wanted, while balancing the largest deal in its history on ephemeral cost-cutting opportunities. They would still be getting Alstom's large installed base of gas and coal power plants, which GE thought it could run more profitably. The deal also included other new capabilities, like the power plant construction business that Alstom maintained, an area that GE had long avoided (in part because it could subject the company to a morass of expensive change orders and political entanglements). But Smith and the others in business development were becoming increasingly alarmed by the growing cost of some of the concessions. Already firmly convinced that GE had entered its negotiations at far too high a price, these executives now recoiled at the cost of the concessions that the European Commission and the Justice Department wanted to exact.

Bankers were starting to mutter too. "Put it this way," one told a *Wall Street Journal* reporter. "Which would have been applauded more — going to do a $17 billion deal for Alstom, or going to do a disciplined divestiture program and shrink [GE Capital] more aggressively? I'm not sure that the Alstom strategy wins out."

Tensions rose, and some on the deals team began to question the motivations of the big law firms that GE had hired to represent the com-

pany on competition and regulatory matters in Europe and the United States. Were the lawyers intent on holding the line and defending the business interests of their client, these executives wondered, or were they crafting whatever deal the regulators required?

Smith and others were even more startled by the proposed solution of Vestager's negotiators to the problem of competition in the turbine market. Alstom was far behind its rivals in bringing a new state-of-the-art heavy-duty gas turbine to market to compete with Siemens or GE's own H-class unit (which itself had been rushed to market after GE fell behind Siemens). Vestager couldn't allow GE to take over Alstom's development program for its new turbine, absorbing what good technology the French company had come up with and discarding what it didn't need, her team said. A vibrant European market required that the turbine program remain in the hands of another European company, and the EC team had decided that it should go to Ansaldo.

That notion didn't necessarily scare GE from a commercial standpoint, as the Alstom turbine was years away from being ready for the marketplace and Ansaldo was a small-fry rival compared to the more formidable Siemens and Hitachi. But Ansaldo wasn't just an Italian rival. The company was now 40 percent owned by Shanghai Electric Company, the state-sponsored Chinese giant.

The global power business was already in a state of permanent, conflicted anxiety about China. With the sluggish global economy, everyone needed China's massive market if they wanted to be able to show sales growth. But the theft of intellectual property in China was so real a threat that GE regularly reassured investors and journalists that it never sold its most state-of-the-art machines there, knowing that if it did they might be reverse-engineered by Chinese competitors. Now Brussels was asking GE to give a turbine program — one less advanced than GE's or Siemens's but better than Shanghai Electric's own program — to a Chinese entity that could be relied on to quickly develop a cheaper competitor for the H-class that would eat into GE's market share over time.

Some at Power couldn't believe the company would even consider this proposal, and yet the deal plowed forward, with the momentum of a corporate strategy set in motion many layers above the negotiators themselves.

One evening early that summer, Smith and deals team members were

once again at work on their models, reallocating values to Alstom physical assets, to their anticipated cost synergies, and to the concessions that GE would need to make to get the deal through. GE executives, already weary of Alstom questions, virtually sighed on earnings calls and at investor conferences when asked the inevitable questions about when the deal would finally close, forcing them to consider the implicit, harsher question of whether this whole exercise was still worth it.

The small team gathered that night was paradoxically delighted when they finished crunching the numbers. By their count, the deal concessions GE was being asked to make had now grown so large that GE's original logic for the transaction no longer made sense. The cost to close the deal seemed to exceed the overall benefit to the company. And that, they realized, would trigger a breakup provision in the sale agreement, allowing GE to back out of what they increasingly viewed as an albatross of a deal.

They weren't the only ones. The company had even gone so far as to hire litigation counsel, anticipating that they would be in for a massive court fight, whether battling over the regulators' demands or finally backing out of the deal altogether.

Adam Smith grabbed his cell phone and called a superior, one of the company officers at corporate who was overseeing the deal. This could be it, Smith told him. We might be able to pull the plug on this thing.

The other executive was firm: "This is Jeff's deal. We're not backing away."

40

CLOSING COSTS

IN THE YEARS to come, Immelt, Bornstein, Bolze, and some board members who backed the Alstom deal would view its failure as a function of poor timing. They'd say that no one could have foreseen how quickly the world would swing from believing in a steady upward trajectory for natural gas–fueled electric power to focusing instead on renewables and battery storage. They'd point fingers, mostly in private, at one another: the Power business had been mismanaged in other ways; corporate hadn't moved quickly enough on synergies; by the time the companies finally managed to get the deal through Brussels in the fall of 2015, the shell of a company they had tried to buy in the spring of 2014 had dried up into a carcass.

But more than anything, they rejected the criticism that had finally begun to leak out into journalists' accounts and the public domain. If anyone had thought this deal was a dog, he or she should have said something, Immelt felt. Bornstein too would insist that the deal had undergone stringent review, and that any GE officer who had objected to it had been duty-bound to speak up. No one had.

That silence was belied by Jeff Immelt's 2015 message to senior staff at corporate and GE Power reminding them that they had all voted to proceed with the transaction, and that Bolze and the Power leadership in particular had signed up to achieve its lofty goals. "It was a love-it-or-leave-it email," one former GE Power executive said. "It [the deal] was to be his legacy."

But the absence of robust opposition also pointed to the broader problem, long cultivated and growing into a quiet crisis within the company, of real candor and self-awareness. For all its vaunted levels of management and procedures of accountability, when it had come time for lower levels of management to stand up to the ultimate boss, Immelt, and tell him that his legacy play wasn't going to work — and in fact, had been a clumsy mistake all along — no one was willing to do so.

For good reason, said Adam Smith (the pseudonym we've given this deals team member to protect his identity). Were they breaking any laws? No. Were they doing something unethical, like putting out a jet engine with a flawed part? Certainly not. So what good would it do to submarine one's career at the company by throwing a wrench into the works of a deal that was precious to the current CEO, and potentially to the man who would succeed him, in full knowledge that the leaders would press ahead anyhow? Having registered his doubts with peers and his superiors — GE was overpaying, and the strategy looked all wrong — Smith ultimately went no further.

In the end, this was GE. How bad could it really be? Mistakes, even big ones, were always absorbed by its size. No one thought it was possible for the bad deals, mistimed decisions, and misallocated spending to pile up high enough to overcome the other successes and strengths of the company. GE was simply too big to be undone by a single bad deal.

The end came on Thursday, September 8, 2015, fully seventy-one weeks since the companies had first agreed on their terms: the European Commission agreed to let General Electric buy the power assets of Alstom SA. The US Justice Department issued its own blessing as well, though, like the EC, it insisted on concessions.

GE agreed to divest PSM, the Florida-based parts and service supplier to Ansaldo, the Italian competitor, preserving competition for power plant owners in the United States, but costing GE a unit that its advisers had thought could be a profit center. In Europe, bowing to Margrethe Vestager's demand that it keep Alstom's turbine program in European hands, GE turned that technology over to Ansaldo as well — virtually ensuring that its technology would eventually reach Shanghai Electric.

For Smith and the deals team, there was one final adjustment to make as the paperwork was finalized on Immelt's transformative deal. The ex-

ecutives tacked on a last-minute adjustment in the purchase price that swung hundreds of millions in Alstom's favor. The French company had an outstanding bill to pay: that massive fine for foreign corruption it had promised to the US government, which had been waiting on its payment pending the closure of the GE deal. All the companies and government lawyers involved had insisted that Alstom pay on its own for the bribery fines, rather than rely on its new American parent to cover the cost.

The price tweak, framed internally as a "brand value adjustment," quietly took care of the problem.

AN UNEXPECTED GUEST

THE CONNECTICUT HEADQUARTERS of General Electric was just off a woodsy freeway; the somewhat sparse entrance, manicured lawn grass and little else, was out of sight from the compound in the distance. It looked similar to a government installation except for a surprisingly small sign in the grass: the GE meatball.

Driving down a curving road, a visitor would come to an all-white, gated guard house.

One Sunday morning in October 2015, Immelt was at work on the top floor of the headquarters with Bornstein and a small group of guests, preparing to unleash GE's next move on the world the following morning. The headquarters was cold and large, both inside and out, and the hallways were a cavernous maze. One executive joked that if he dropped dead in one of the hallways, it might be days before his body was discovered. On some days when few people were around, Immelt could be heard blasting music from his office — the Allman Brothers or other guitar rock of the 1970s — as he worked. Some executives say that these were the times they connected best with Immelt, when he was informal and lowered his guard.

There was no music blasting that day. Some of the guests were from a new major investor in GE, Trian Fund Management, an activist firm usually feared rather than embraced by top executives. The other guests were *Wall Street Journal* reporters, who were there to learn about the arrangement, interview the CEOs, and drop the news the next morning.

The dance had begun months earlier, after GE's announcement that it would be selling most of the GE Capital divisions. For many investors, this was an enormous step in the right direction. Nelson Peltz, who ran Trian, called to congratulate Immelt on the move right after the news became public.

In response, the GE boss told Peltz: "We'd love to have you in the stock."

Some outsiders saw Immelt's invitation to a disruptive investor like Peltz as a sign of confidence, but it was also a defensive strategy. Other activists were circling GE, casting a shadow over the transformation agenda.

Immelt knew that one of the big risks was an outsider coming in to try to disrupt the plan. Even though GE was a huge company and it would be difficult for anyone to amass a sizable position without significant capital, even large companies had increasingly become the target of activist shareholders, and conglomerates were the dream target. Activist firms loved to take companies apart to find more value in the pieces than the combined operation, and they often went after entrenched CEOs who might not want to loosen their grip on power, using their ownership stake to unseat boards and force strategy changes. But one thing was fairly certain about activists: they didn't often invest next to each other. If one took a position, others were unlikely to follow.

Regardless of its insistence that it was merely a "highly engaged shareholder," Trian was undoubtedly a major activist in the market. But to Immelt, it was at least a familiar one. Peltz had come to speak to GE executives in the past, and Trian cofounder Ed Garden had known Immelt since he was a teenager; his older brother was in the same Dartmouth fraternity with Jeff, and he'd listened to Immelt speak at the school's graduation. For Immelt, these connections made Trian's involvement seem less threatening, though he knew the risks and knew that Trian wasn't going to shift its own strategy for GE. But Immelt had convinced them to endorse his plans for GE, assuring them that Trian would benefit from the changes at the company that would inevitably lift the stock price.

Trian wasn't calling for a breakup, as it had at other lumbering companies, including DuPont and Kraft. Trian wasn't even seeking a seat on GE's board, whose eighteen high-powered members were loyal to Im-

melt, as it had done at Family Dollar, Ingersoll-Rand, Mondelez International, and PepsiCo. Rather, the influential hedge fund was coming forward with what amounted to a high-profile endorsement of Immelt's strategy, via a massive $2.5 billion investment in GE.

Garden was chief investment officer and also CEO Peltz's son-in-law, a fact that he didn't love having pointed out in the press, because he felt it undermined his importance to the firm. Garden had cofounded the fund and was the likely future head of operations. By some claims, Garden did much of the heavy lifting at the firm. If he were to run it for decades to come, having investors think that Peltz was the key to its success might cause some serious problems down the road.

But Peltz and Garden were in sync when it came to running companies. They pushed for efficiency and smart capital allocation, and they looked to cut costs. The coffee mugs in their wood-paneled New York office bore the motto CASH IS KING. When it came to GE, the two Trian leaders made it clear that their investment would be a partnership with management.

They had come to Fairfield that Sunday to explain their thinking. As the meeting began, Immelt was wedged in a chair next to Bornstein, whose body language belied the kumbaya atmosphere that the communication staff was working hard to manufacture. Bornstein often wore a scowl, but it was the rest of his posture that gave him away this day. His arms folded across his chest, the CFO never quite turned in his chair enough to look directly at the self-assured activists who were about to reveal that they'd bought a giant stake in his company.

While Garden and Peltz wore sharp tailored suits befitting men who controlled billions of dollars in investments, they also wore sneakers. That note of informality amplified their power as they sat in the inner sanctum of GE's board, the place where Immelt held almost absolute power. Amid the fine art displayed in the room, the sneakers were a reminder of their sovereignty.

"It's not something you want to break up," Peltz said of GE, sitting in the wood-paneled boardroom. "It's something you want to keep taking care of."

Trian had done its homework before amassing its position over the previous month. It laid out its findings in an eighty-page white paper titled "Transformation Underway . . . But Nobody Cares." The firm ar-

gued that GE's stock, then trading around $25, could reach $40 to $45 a share by the end of 2017.

The white paper praised the Alstom deal and, in a proposal that wouldn't age well, pushed GE to borrow more so that it could repurchase another $20 billion in stock. Trian, as always, had to rely on publicly available information in its analysis, a fact that it would point to later when explaining its conclusions and recommendations.

Peltz was the high-profile name. A friend of Donald Trump's, Peltz had dropped out of Wharton as an undergraduate and turned his family's frozen food and produce business into a runaway success and public company. His first major success as an activist came with the turnaround of Snapple, the beverage company.

Garden, a lean man who wore clear-rim eyeglasses, was the calmer side of the partnership. He had little problem speaking his mind, however, and consistently hammered the message that Trian helped fix companies; it wasn't one of the corporate raiders that so many CEOs feared. Trian was a force for positive change, he insisted. That said, they had successfully broken up many companies too, when they felt their flaws couldn't be fixed.

Trian's endorsement was the stamp of approval that Immelt thought would help others realize the full legitimacy of GE's expected turnaround. It was an inoculation against some other more abrasive investor trying to buy a stake and change the strategic direction — or perhaps even oust Immelt from the chairman's seat before his masterstroke had the time to succeed.

By the end of 2015, with Trian's blessing, Immelt's plan was coming together. The Alstom deal was closed, the sell-off of GE Capital was going faster than expected, the marketing push for Digital continued, and the stage was set for GE's next era of greatness.

42

THE CANDY FACTORY

THE STAGE WAS set, but the next chapter of the General Electric story wasn't going to be written from the woods of Connecticut. After four decades on a sprawling corporate campus, the company that presented itself as a corporate titan on the leading edge of business — all tied together by its massive research organization and its much-hyped software capability with Predix — wanted to pull up roots for a more appropriately urban setting.

GE couched its departure from Fairfield as a reaction to an oppressive liberal state government. Reaching out to a local lawmaker through his wife, a Republican officeholder in Ridgefield, Connecticut, where they lived, Bornstein threatened that GE would consider leaving the state if a proposed state budget deal included some business and property taxes that had been approved by a majority of lawmakers. In fact, GE's departure was already a fait accompli. The company leaders already wanted to move to a city, something their corporate peers had been doing for years. Bornstein himself thought of Fairfield as "a morgue." The company's lobbyists hadn't pressed for major changes to the tax hikes under consideration; their primary ask had been to preserve GE's ability to use the billions of dollars of carried-forward tax losses that it used to cut its tax rate, generated by GE Capital, in perpetuity. When budget officials from Governor Dan Malloy's administration offered to preserve the losses in exchange for a pledge to stay in the state, the company demurred.

They were also settling older scores. GE leaders, including Immelt and Bornstein, had been incensed when Connecticut's federal delegation worked to preserve Pratt & Whitney's role as the exclusive engine supplier for the new F-35 joint strike fighter. The lawmakers had taken Pratt's side by blocking GE's attempt to offer the Pentagon a second engine option, which had the potential to provide a huge new stream of business for GE Aviation. For the representatives and senators, the decision had been a no-brainer: Pratt & Whitney and its parent, United Technologies Corporation, were the largest private employers in the state. The engines would be built by union workers in the company's massive factories in East Hartford and Middletown. GE had a headquarters and many wealthy executives scattered around Fairfield County, but nothing of the importance, economically or politically, of a giant cohort of middle-class machinists who voted.

The final straw had been a victory celebration thrown in East Hartford by the Connecticut delegation, a thumb in GE's eye after the second engine proposal had been killed for good. Malloy, just days into his term as Connecticut's first Democratic governor in a generation, accepted the congressional delegation's invitation to join the celebration. That, a GE lobbyist said, sealed his fate. Some things the company did not forgive.

Now openly courting a new home, GE used its selection process as a political nightstick. GE had spent considerable effort lobbying Congress to reopen the Export-Import Bank, which provides financing to overseas customers of US manufacturers. In the headquarters search, GE publicly removed several candidate sites from consideration because their congressional representatives were opposed to reauthorizing Ex-Im, which GE viewed as essential, but which had been vilified by some on the left and right as unnecessary corporate welfare.

After a high-profile bake-off among various cities trying to lure the iconic company, GE announced it would move to Boston.

The company rolled out plans to restore two former candy factory buildings and build a cutting-edge — if terribly ugly — modern glass tower in Boston. GE also scooped up $145 million in incentives for the move from the city and state governments and boasted to investors that with the incentives, its move would ultimately cost the company nothing.

The move to Boston was filled with photo ops and the expected

swooning by the local press, bedazzled by one of the world's biggest companies seeming to just fall into their laps. GE pushed hard on its messaging around the digital business and the talent it would need to hire in the Boston area for future endeavors. Like the Alstom deal, GE made a commitment to bringing new jobs to Boston.

Immelt now had all the pieces in place for his grand vision. The industrial businesses were positioned to grow enough to completely disconnect from GE Capital. Cash from selling financial services would be used to buy back GE shares, perhaps Immelt's favorite use of cash. The reduced number of shares would boost earnings per share and cut the total cost of the quarterly dividend — offsetting the loss of GE Capital.

These were the central tenets of the pivot that would bring about the company's transformation — a simple, logical plan built out of a matrix of moving parts. Now it was just about executing the plan. And operational excellence was supposed to be synonymous with GE.

"In a complicated world, we are simpler and more competitive. In an uncertain world, we are skilled in managing through tough cycles. In a risky world, we have cultural strength and a lot of cash," Immelt wrote to shareholders in early 2016. That long discourse, delivered in Immelt's familiar tone, was essentially a lecture on how to perceive the company, saturated with the CEO's trademark unapologetic optimism and plenty of marketing pitches.

In many ways, Immelt was feeling good. At the end of 2015, GE's stock finally crossed $30 for the first time in seven and a half years. With the white whale within his sights, perhaps the Trian endorsement had been the chum in the water that Immelt needed to get investors excited about the stock. Once they saw the vision coming together, they would recognize the brilliance of the company's positioning. And now the stock was — finally — moving in the right direction.

Unfortunately for Immelt — and the legions who held GE stock options that continued to sit untouched — the stock would have a hard time going much further. The frustration of wallowing below the $30 benchmark deepened with the realization of being stuck there.

In the annual letter, Immelt wrapped up his lecture on the limitless superlatives of GE with an awkward public plea to major institutional investors — a blanket term for banks, mutual funds, hedge funds, pension funds, insurance companies, and other large investors. Institutional

investors own the vast majority of the equity markets, and they aren't usually pitched in public by the CEO of a major corporation.

"We have delivered for you in the last five years. But we are still under-owned by big investors. In this time of uncertainty, why not GE? We have great businesses, global scale and strong initiatives," he wrote, like a heartbroken lover begging for reconciliation. "We have a ton of cash that can protect you," he added. "And we will lead the Industrial Internet. We are the Digital Industrial. We have grit. Our leaders learned from the experience of economic volatility."

For the Power business, it was pedal to the metal. It was launching its next-generation natural gas turbine — in essence, a cousin of GE's jet engines, but vastly larger, and capable of turning an equally large generator to produce electricity. The newest machine was air-cooled and operated at temperatures higher than the melting point of the materials of which it was made. No one questioned that it was a marvel of industrial engineering.

The company's swagger about its Power unit was hard-earned and long-standing. GE had built the first natural gas turbine to run a power plant in the United States in 1949. The first unit, which pumped out even more electricity than the planners had hoped for, wouldn't be decommissioned until 1980. The first gas turbine ever was run a decade earlier in Switzerland by a company that, through a series of acquisitions, ended up being acquired by Alstom in 2000. Now it too was part of GE.

If the Alstom deal was a crucial piece of Immelt's plan for success, it was a dream come true for Steve Bolze. Immelt's devotion to the Alstom deal was Bolze's chance to step up to the plate. He had been running the Power division since 2008, his star rising when Vice Chair John Krenicki abruptly left the company in 2012 at age fifty, with a payday that included $89,000 a month for ten years, after which his pension plan would kick in. (Krenicki, insiders said, had been forced out by Immelt, in part because he had already seemed a little too prone to disagreeing with the CEO or telling him no.)

With the Alstom deal done, the mission was clear. Jeff Immelt wanted to corner the market in natural gas turbines. If Bolze's relationship with Krenicki had been rough, he was now heading to the same place, as he would butt heads with Immelt over similar questions of targets and

earnings. But he didn't think that mattered and Bolze still thought the top job would eventually be his to lose. He wouldn't have still been steering the ship at Power if that weren't the case.

As 2016 opened, GE finally had Alstom securely under its umbrella and was working to integrate the new operations. In March, Bolze traveled to New York to present at an investor conference in the JPMorgan offices, where he would be questioned by the bank analyst who covered GE.

The JPMorgan analyst, Steve Tusa, liked to start trouble. And he wasn't particularly fond of Bolze.

YOUR PEOPLE DON'T WANT
IT BAD ENOUGH

U NBEKNOWNST TO BOLZE, Tusa was preparing to drop a bomb on
GE and its investors with the first "sell" rating on the company by a
major bank's analyst in recent memory. Given GE's history with JPMorgan, from its historic founding through advising on the rapid exit from
much of Capital, the vote of disapproval was a stinging slap.

Months of examining GE with his team had led Tusa to conclude that
the company was overvalued and its outlook wasn't as rosy as it presented. At the Power business, the core of the company, he detected that
the aggressive use of upgrade packages was increasing sales, but not in a
way that would be beneficial over the long term.

In essence, GE was selling upgrades to existing customers on the
promise that it meant less maintenance and more efficiency. The problem was that the future maintenance would also be GE business, so in
effect the company was making a sale today that would eliminate work
to be done (and billed) tomorrow.

In finance parlance, GE was pulling sales forward. It was agreeing to
pay Tuesday for a hamburger today. Not knowing what the analyst had
concluded, Bolze approached the session with Tusa with his usual confident manner. He pitched the benefits of the Alstom deal and explained
the importance of Predix, speaking about it as if most power customers
were using the software and were now getting benefits.

"What is Predix? Predix is cloud-based operating systems that our
customers can put on their equipment that allows them to more effi-

ciently capture all the data on that equipment," Bolze recited, noting that GE's customers were increasingly eager to have the system.

Deeply entrenched in the GE view that the world was heading into the "Age of Gas" for decades to come, Bolze then made an assertion that would fall flat. It also gave a glimpse of GE's blurred view of the market that it planned to corner.

"The world needs about 50 percent more power in the next two decades," he said. "The biggest power source going into the next decade is gas . . . Gas will grow almost 50 percent more in the next decade, and we are positioned very well there in that space."

Days later, Joe Kaeser, CEO of Siemens, would offer a different view. Kaeser was perhaps Immelt's biggest rival. He looked a bit like Bolze, but with European styling (spread collars, modern eyeglasses) and hair that was more salt than pepper. The German conglomerate competed in a lot of the same markets as GE, and the two CEOs had an uneasy relationship. Kaeser had, of course, taken a run at Alstom with Mitsubishi and would later criticize GE's cut-rate pricing in the power market, which would exacerbate an already tense environment.

To Immelt and his followers, Siemens's interest in Alstom was proof of why the French company needed to be owned by GE rather than competitors. From Siemens's viewpoint, they had succeeded in dragging out the Alstom process and pushing up the price for GE.

Unlike Immelt, Kaeser wasn't sold on the conglomerate model and was moving Siemens in the opposite direction and breaking up the company, although he would never describe it in that way. But the outcome was the same. Siemens was slowly but surely separating its businesses and keeping a stake in the new entities under a strategy Kaeser called "Fleet of Ships."

When it came to the power market that Bolze saw as ripe for harvesting, Kaeser conceded that there would be growth in the very large natural gas turbines that Bolze was touting. But he gave an ominous warning.

"So there will be growth . . . But at a cost, because every order is very, very competitive," he said.

For Bolze, the path to success was like a rope bridge over a deep chasm — the only way to get to the other side. The summer before the Alstom

deal closed, Bolze was giving another presentation, this time in the auditorium at Crotonville. Behind a desk, watching his every move, was Immelt.

When he concentrated, Immelt scowled slightly in a way that was almost a squint. He tilted his head back and looked down his nose. Bolze cued up his PowerPoint slides, the lightsaber of the GE universe, for his slot in the Growth Playbook, a grueling annual examination of GE's eight major business leaders.

It was here that GE hammered out targets for sales and profit, setting the underlying assumptions for the financial estimates it would give investors. Under Immelt, the point of the exercise was determining how his executives would get to their financial targets — though not how they would determine what output the businesses would produce as a starting point. This practice had been ingrained at GE from the days of Welch, was continuing under Immelt, and presumably would roll on into the beyond.

It had been a few months since the plan to sell off GE Capital had been announced, and the Alstom deal was creeping toward completion. Just as he was at the JPMorgan conference, Bolze was confident and presented himself as the GE *Übermensch*. His face, résumé, and demeanor all declared that he was destined to run GE one day.

Bolze's forthcoming global portfolio of new power plants and thousands of workers hired seemed to ensure that fate — as long as there were no major surprises. Moving through the slides, Bolze came to the proposed annual sales growth rate of the Power business: 5 percent.

It is difficult to put that prediction into context, even in the endlessly optimistic environment of GE. Suffice to say there was ample reason for skepticism. Power had been struggling to meet targets, and its sales hadn't grown that quickly in years.

The reality was that global investment in new gas-fired power plants was slowing and promised to be lumpy at best. Energy efficiency was on the rise, and concern about the warming climate had raised real questions about the future of electricity produced by fossil fuels. Natural gas was cleaner than most fuels, and new turbines were very efficient — but there was still air pollution spewing out of those power plants, and the cost of renewable alternatives was steadily falling.

This pressure meant that future GE revenue from the highly profit-

able service contracts it had signed was likely to fall, or at least would not grow as quickly as predicted. Global gross domestic product, a reliable proxy for the power market, was below 4 percent.

The target Bolze gave Immelt was based on a rosy assumption that clearly cried out for interrogation — which, after all, was the point of having the formal review.

As the room watched, Immelt gave the desk in front of him a confident slap.

"Great, next page," he said. He knew that the Power business was crucial to getting him to his precious $2 per share target for 2018. Bolze had to come through, and it really didn't matter how. Immelt would make sure of it.

Immelt could be tough on executives in his own way in these briefings, but not usually for being too optimistic. "Where's the guy I used to know?" he would ask an underling who told him that Immelt's targets couldn't be hit. Immelt's locker-room chumminess and absolute power at the company made such comments sting. And even if it was never delivered explicitly, the lesson sunk in, down through the levels of the company. There was no market for hard truths or bad news. Not as far as the guy at the top was concerned.

A targeted executive never forgot this kind of rebuke from Immelt. It was better to figure out a better way to deliver the bad news, or make it go away somehow, than to present it to him straight.

When Immelt's disappointment soured the mood, the tone changed. "Your people," Immelt would say, "don't want it bad enough."

Executives would stretch far to avoid that tone. And that is exactly what they did in Power.

44

MAKING ADJUSTMENTS

JUST AS KAESER predicted, Immelt pushed for market share at all costs. To win deals, GE's sales force did anything it could, including cutting deals on more and more aggressive terms, up to the point that the contracts they signed might be unlikely to turn a long-term profit at all. And as Immelt had envisioned, salespeople relied on financing provided by the stub of GE Capital to prop up customer demand.

In Immelt's thinking, if he could capture the market now, it barely mattered how he did it, because tomorrow he would own the world. The Age of Gas would belong to GE.

But the pressure from renewable energy sources was already rising. While governments were offering piles of incentives for new power generation from sources like wind, solar, and water, GE was still trying to weave together the Alstom operations in the biggest such combination in its history. GE was attempting to absorb and integrate a behemoth of the old energy economy, not the new.

Facing the slowdown in fossil fuel equipment sales and competition from renewable energy, GE Power managers turned to an old trick, one long in use in the Aviation division. Jet engines are generally sold at a loss, and the real money is made in the decades ahead of servicing them. When selling such equipment, GE usually signed up a customer for a long-term service contract. The company used that same long-range service model in the Power business.

The accounting rules for service contracts give leeway to companies

to estimate the profits they will produce in future years. Importantly, when a company adjusts those estimates of expected future income, it records the changes in the current accounting period. Thus, if a company can consistently find ways to ensure that a long-term contract will be more profitable in the future, the contract can be a source of income in the present. When the contracts are long, some even decades out, tiny tweaks to assumptions — the price of a part, the schedule of maintenance shutdown — can produce huge changes in the end result. A rosier estimate of a contract's value in the future can allow companies to record a profit — on paper — today.

Those changes aren't supposed to be made arbitrarily, or to manufacture profits. Companies are expected to alter their contract profitability assumptions only when they are reasonably sure that the equation has changed. Being wrong about that — moving assumptions about future revenue too far in the wrong direction — can cause quite a headache and bring losses as the contract runs out its life.

For GE, a technological innovation that improved the performance of a turbine blade or lengthened the number of hours between maintenance outages had to be accounted for in its portfolios of service contracts. And much of the company's research and innovation was aimed at improving the long-term profitability of those contracts. Predix, for example, was touted as improving the operation of massive machines, meaning that just these types of profitable efficiencies were virtually guaranteed to materialize if the software worked as advertised. As always, GE sought to ensure that the cost of fulfilling its contract promises would decline over time, and that meant more profit.

Throughout 2016, Power teams combed through the division's portfolio of service contracts, each representing payments from power generators to maintain the turbines that GE had sold them. One strategy they often used was to offer discounted turbine upgrades to customers in exchange for extending the length of their contracts as far out as 2050. Such long contracts allowed for huge payoffs — paper profits — once the assumptions were adjusted.

Executives also scoured existing contracts for ways to change underlying assumptions. The entire book of GE's service contracts was reviewed at least once a year.

These reviews produced profits that GE could use to hit targets for

Wall Street, but they were really future profits, produced by accounting adjustments alone. There was no actual new cash coming in. Such gaps between reported income and cash flow usually close pretty quickly, because they represent a simple timing issue — a bill wasn't paid, or cycles didn't match up to quarters. When such gaps are frequent and not closed quickly, it can be a red flag to investors. They see a telltale sign of aggressive accounting, if not fraud, when a company records a lot of profits but no cash to back them up. Anyone can run up a credit card, but someone has to pay that bill eventually. While contract accounting is complicated, investors still expect to see real money coming in the door.

Doing its part to meet GE's earnings targets, the Power team was finding just these kinds of paper profits in the vast portfolios of service contracts it maintained on gas and steam turbines around the world. But they were also digging a future hole — pulling forward the value of those long-term assets, on very optimistic terms.

To pad the hole, GE now began selling its receivables — bills its customers owed over time — to GE Capital in order to generate short-term cash flow, making it appear that those newfound profits were matched by cash flowing in the door. But a future invoice can be sold only once. GE Power was making its numbers in a way that would make future growth more difficult. This accounting maneuver was legal, but aggressive. In time, that move would catch the attention of criminal investigators at the US Department of Justice.

Bolze's team was operating in a tradition that stretched at least as far back as the glory days of Jack Welch, when tweaking service contracts was a tool occasionally used. Those who worked in Aviation at that time have described their awareness of the danger of being too aggressive with contract adjustments. Taking those noncash gains could create long-term problems that outweighed any benefit of a quick boost in earnings. Now, however, service contracts were being tweaked with less caution. The scale of the ambitious contract accounting was far bigger, and as Jeff Immelt's pivot continued, the stakes of Power continuing to hit its numbers were higher than ever.

By the end of 2016, worry was starting to grow inside Power. Management's expectations about the sales growth and profit they should be able to hit didn't reflect the dim reality of the market, team mem-

bers told Steve Bolze and Paul McElhinney, the head of the unit that administered the service contracts. McElhinney, a gray-haired Irishman with a long pointy nose, had held a similar role overseeing services in the Aviation division and had brought the same aggressive approach to Power.

Vocal complaints about management's view diverging from the reality of the market, or from basic math, were common among lower-level Power executives. When these concerns were raised to leaders like McElhinney, they were stopped cold.

"Steve's our guy," McElhinney said in one meeting. If Bolze was elevated to CEO, those behind him in Power would rise too. If there were problems growing in the service division's numbers, they could worry about a solution when they arose.

"Get on board," McElhinney said. "We have to make the numbers."

45

BACK BAY DEAL

THE FEELING OF fragile détente between GE and Trian that had filled the boardroom a year earlier didn't last long. The hedge fund had bought into Immelt's investment thesis that GE was changing its structure and that it was only a matter of time before the market woke up to the company's transformation. But after years of being categorized as a diversified financial services company rather than a pure industrial — a branding that Welch resisted — it wasn't so easy for GE or its shareholders to change gears.

For Trian, GE was an investment, not a partnership. They did not have the same approach as the GE board, which was seemingly content to swallow whatever explanation it was fed by Immelt. It didn't matter that Immelt and Ed Garden had known each other decades earlier.

Garden knew that Immelt was smart. He'd gone to Harvard Business School and had the familiar marks of achievement that all of the members of this fraternity of corporate success tended to share. But Garden trusted no one — he wanted to see results, and he wasn't one to make the same mistake twice.

Immelt had already been given the benefit of the doubt when Trian tied up billions of dollars in GE stock and claimed no seat on the board in exchange, as it usually would have. Trian was clear from the start, however, that it would be keeping a close eye on its investment.

Moreover, Trian had its own investors. They expected results, and this was the biggest bet the fund had ever made.

A year after the initial Trian investment, GE was behind on its financial targets and the stock wasn't moving. Garden and Trian viewed the missed targets as not just disappointing but insulting. Garden took it personally. Trian's big public endorsement had boosted Immelt and suggested that the fund would remain an investor in the stock for years. For GE to fail to hold up its end of the bargain in the first year was shocking.

In the fall of 2016, an increasingly impatient Garden traveled to Boston to see Jeff Bornstein at his six-level $13 million townhouse in Boston's Back Bay. Like many on Wall Street, Garden saw Bornstein as more of a straight shooter than Immelt. Bornstein had come up through the world of finance. The two men spoke the same language. Immelt had been well served by his ability to spin inspiring visions of the future, but Trian didn't want a politician or a seer right now. Garden wanted the cold-hearted calculation of a banker.

Visiting Bornstein, Garden made it clear that if GE's performance didn't improve, Trian would ask for a seat on the board. The two men got along, but there was also an unspoken tension between them. Garden didn't shy from a fight, and as an activist, he was comfortable with the implicit confrontation that his very presence represented.

Bornstein had a similar view. He was at the top of the financial structure of GE, which was its own kingdom. His decisions could move markets, and he had been central to steering GE Capital through the crisis. He really only answered to Immelt, and who knew? He might even run the place someday.

As the meeting went on, Garden's complaints didn't sit well with Bornstein. GE's size plus its influence in Washington and around the world gave it tentacles into all corners of society. That meant that it didn't have to bend to anyone. And company lifers like Bornstein were sure to their bones that they knew better how to manage a complex beast like GE than any shortsighted activist. Bornstein was also deeply wary of the tactics that Garden and Peltz wanted GE to use to goose the stock price, especially taking on even more debt to buy back more shares. "If they want to do that," he muttered at one point, "they'll have to find a new CFO."

But Trian was willing to stage a public battle, which GE wanted to avoid. Even if Bornstein was up for the challenge, Immelt wouldn't have it. An activist fight could send Immelt into retirement, leaving him on

the outside while the shoots of GE's rebirth were just breaking through the soil.

GE's fear of a public fight gave Trian just the leverage it needed.

Bornstein and Garden started to work out a compromise. GE doubled its cost-cutting goals and tied more of its executive bonuses to profits in its core industrial units, which was where Trian wanted the company to focus. Having hard numbers for targets also eliminated any debate on how to measure success.

Some parts of the agreement weren't made public. But the bottom line was that if GE didn't get back on track, Trian would push for a board seat or management changes or both. Both men understood that "management changes" meant defenestrations, including of Bornstein himself, even if many within the company thought that the CFO still was on track to succeed Immelt.

46

MANAGING POWER

J EFF IMMELT STILL believed — in a General Electric that was for-
ever, and in an idea of his own position as something closer to a pub-
lic calling than another high-paid corporate sinecure. He and his closest
defenders always emphasized that, when asked by President Obama to
lead the President's Council on Jobs and Competitiveness, he'd felt duty-
bound to accept the task, General Electric being a part of the American
civic fabric as well as the economy. And personally, Immelt had felt pa-
triotically bound to honor his country's request.

But serving on the Jobs Council had been a nightmare for Immelt.
It was all exposure and no results in a still sluggish economy, and his
visibility on the council only heightened the scrutiny of his company
and himself. Internet pundits now declared Immelt, a self-avowed "cen-
ter-right Republican," to be an Obama stooge. And many people, both
within and outside the company, hated the CEO's attention being any-
where other than on the project of reviving GE's lame stock price.

Then the country went and elected one of Immelt's ex–TV perform-
ers president.

Some within GE's upper echelon had shuddered at the election of
Donald Trump. (And some cried.) But many were confident, as their
peers were, that the volatile businessman and TV celebrity would be al-
tered by the job.

After all, they had bent Donald Trump to their own ends before. In

the 1990s, GE's asset management division — the financial unit that made investments on behalf of the company's huge pension funds — had foreclosed on a defaulted loan, taking ownership of the creaky old Gulf and Western Building overlooking Columbus Circle, the gateway to New York's Central Park. The asset division had solicited bids to redevelop the dated structure from a slew of major New York developers, including Larry Silverstein and the Rudin family. They had also agreed to accept a bid from Trump, then at a low ebb brought on by serial casino bankruptcies and baroque tabloid coverage of his messy sex life.

Despite their doubts about Trump, the pension plan executives eventually decided that they liked Trump's plan for renovating the office building into hotels and apartments and could make use of his notoriety and brand in getting the units to sell. "Jack was not exactly excited when I brought him that deal," one said later. The executives had asked their external counsel, Sandy Morhouse of Dewey Ballantine, to lay out to Welch in writing the reasons that doing business with Donald Trump was worth the reputational risk to GE.

But the deal had worked, and what all sides remembered now was that they had made money.

Some GE executives chuckled years later at how Trump took credit, in his books and public remarks, for aspects of the development that they felt he'd had little to do with, like installing the high-end Jean-Georges restaurant in the tower as an anchor tenant. Still, at least one aspect of the development had been Trump's brainchild, they noted: setting up the building as a condo-hotel, then an unusual arrangement in Manhattan, but an effective end run around local zoning.

Now Immelt, for his part, showed up at the White House. He laughed and chuckled among a group of CEOs at a gathering in the State Dining Room, seated across the room from the president, next to Vice President Mike Pence. Immelt had known Trump ever since they had launched *The Apprentice* together on NBC, the network then owned by GE. That show had rehabilitated the famously washed-up developer's image, introducing him to a new generation of Americans not as a joke but as an apparent success.

Trump went around the long rectangular arrangement of tables, encouraging the corporate titans to remark on their excitement about his

election and hopes for the future of the economy under a Trump administration. Immelt waited patiently for his turn to respond as a horde of reporters and cameras took it all in behind a velvet rope.

"Jeff actually watched me make a hole-in-one," Trump said, arriving at Immelt's place after receiving encomiums from the CEOs of Campbell's Soup and Lockheed Martin. "Should you tell that story?"

Big Jeff obliged. Chuckling obsequiously, he related the story of Trump's lucky eagle, a shot struck on the golf course as Immelt and NBC were laying out the terms of the deal for *The Apprentice*. Immelt delivered the punch line: Trump had proclaimed that he was the richest golfer in the world, the CEO said, to laughter from the crowd in the White House.

Now the president cut in to correct him. "I actually said I'm the best golfer of all the rich people."

His large frame slightly hunched at the table in the opulent dining room, Jeff Immelt blushed and laughed along.

After the election, most companies weren't looking to get noticed by the incoming president, who had already targeted AT&T and Boeing. It wasn't clear how the new White House would treat businesses going forward.

Immelt often reminded people that he was a Republican, despite never quite shaking an association with President Obama after leading his jobs council, but wasn't backing down on GE's strategy or direction because of Donald Trump. Immelt embraced the situational pragmatism of the high-powered. His priorities were the company's — deregulatory and laissez-faire — but ultimately mutable, in response to the needs of the moment, and contested only when the bottom line was truly threatened. That could lead to unexpected divergence from stereotype. For instance, when President George W. Bush rolled out his tax cuts, Immelt told one investor that the plan didn't make sense. "Why cut my taxes?" he said, referring to his wealth. "Why not let the poorer people have it and spend it?"

The company and its leaders could also at times take a diametrically opposed position. GE had been able to point to few concrete impacts on its fortunes when it cited a progressive new budget law as its reason for storming out of Connecticut to move to Boston. (The company's tax as-

sets included enormous stores of financial losses that it used to hold its tax rate down; in most years, sources told the *Journal*, the company's tax bill to the state amounted to the minimum: a filing fee of $250.) But that budget had significantly raised income taxes on the state's rich, a cohort that included a lot of executives at General Electric, including those at the very top.

GE was a market leader: it validated trends and set the tone for other companies. As the quintessential American company, GE's growth was America's growth. And leaders like Immelt felt they didn't run away from problems and that included a Trump presidency. The successful Trump campaign and the administration's early moves were all rooted in a populist mercantilism that was the opposite of the worldview in which Immelt had overhauled the company.

After Trump's threat to pull the United States out of multinational trade deals, Immelt used his annual letter to shareholders in February 2017 to remind them that GE was bigger than any one country.

"We don't need trade deals, because we have a superior global footprint," he wrote. "We see many giving up on globalization; that means more for us." And Immelt didn't shy away when Trump's deregulation agenda conflicted with GE's stance on climate change. "No matter how it unfolds, it doesn't change what GE believes," he wrote in a note to employees in March 2017.

The CEO was no dummy. Carving out distance from Trump, Immelt also carefully avoided criticizing his touchy former TV talent directly.

Amid the turmoil, and despite the pressures to perform, GE was still hunting for big deals. A team in the Aviation division had worked with bankers to put together a proposal to buy aerospace rival Rockwell Collins in late 2016. The deal pitch, worth more than $15 billion, reached Immelt in early 2017.

He scuttled it, to the frustration of those who saw such a deal as being far more lucrative than Alstom. But Immelt knew he was being watched and was determined not to drop the ball. He couldn't afford Trian, or others, seeing yet another example of him seeming to bite off more than he could chew. Instead, GE kept repurchasing stock, spending more than $3 billion in the first four months of 2017.

GE Power, the unit that led all others in sales, was the centerpiece of Immelt's new GE, and that was how he wanted it to stay. His frustra-

tions with Bolze, who sometimes signaled that Immelt's views of the market were aggressive, led him to reach out to Bolze's underlings. Immelt knew the power of his influence, and he wasn't above calling these subordinates to make sure they knew the stakes and to urge them to hit their targets. Bolze got wind of Immelt's meddling, of course, to his annoyance. The relationship between the CEO and the chief of GE's biggest division was splintering.

There were only so many service contracts to be renegotiated. Power was running out of future profits to pull forward. Its mysterious outperformance in a slowing market wasn't going to last much longer. Cracks soon started to show. That April the company revealed the weakness hidden inside the Power unit with a single, startling figure: GE's industrial businesses had burned through $1.6 billion in cash in the first quarter. That was about $1 billion worse than it had projected.

Bornstein insisted that it was just a matter of timing and that Power's financials would rebound later in the year. Nevertheless, the news raised red flags about aggressive accounting and doubts as to whether the company could make its goals. Some saw these developments as little more than the same roller-coaster ride GE took them on every quarter. But others were concerned that it was a bigger problem than that. If Power was so strapped for cash, how could GE hit its $2 per share earnings target, on which Immelt had staked what was left of his credibility on Wall Street?

To close observers, there was a greater concern: Most of the shortfall came from the heart of the industrial business, Power's service contracts. Services, as in any traditional industrial business, should have been the source of the easiest profits. Instead, the cavity in the heart of the service portfolio was about to tip the entire company into crisis.

UNMOVABLE TARGET

MAY 2017 BROUGHT on EPG, the shorthand for the annual Electri-cal Products Group conference at which industrial investors and ex-ecutives gathered. By tradition, GE was the headliner. It had the last slot on the last day. Some said that it had that spot because it was the biggest company there, but others considered GE's placement in the schedule just a ploy to discourage people from leaving the gathering early.

The conglomerate's prospects had looked promising in early 2017, with the stock above $30 and the broader stock market showing strong momentum. GE investors then were hopeful that Immelt's big bets on oil and power generation would finally pay off.

But by May the stock market was booming and GE wasn't. Trad-ing down 11 percent for the year, its stock had dipped again below $30, much to Immelt's ongoing annoyance, and investors were seeking reas-surance that GE's financial targets remained intact. They openly won-dered whether Immelt would stick by his 2018 profit target of $2 a share. Senior executives were perplexed by the long-held target, and Jeff Born-stein, the CFO who had given his word to Trian at the risk of his job, had privately urged Immelt to back away from a promise GE seemed unlikely to keep.

Immelt was as accomplished as anyone at the task of presenting to crowds like EPG's. He had honed his ability to navigate a deck of Power-Point slides over the decades. But this year was different. The confident, affable salesman ready with a smile and a joke just wasn't himself as he

faced a skeptical audience inside the ballroom of the Longboat Key Resort in Sarasota, Florida.

Shakily, he raced through the highlights of his slides. On the last one, he defended the company's 2018 earnings goal. Sort of. If the oil and gas markets didn't improve, Immelt said, the $2 target for 2018 would be a reach and the company would have to cut even more costs to have a chance of delivering on it.

In fact, Immelt knew that in all likelihood his last and biggest pledge to the market was going to come up short. It was wishful thinking at best that GE could deliver the $2 of earnings he had promised. But Immelt didn't say that onstage at EPG.

In the CEO's view, his eventual successor would inevitably have to reset the earnings target. Cutting the target once would be bad enough, surely triggering an angry sell-off of company shares. Better for investors and the company, in Immelt's thinking, to wait and let the next boss decide when and how to take the $2 target down.

Now, though, the crowd buzzed with confusion.

Barclays analyst Scott Davis asked bluntly if Immelt was backing the target.

"It's going to be in the range, Scott," Immelt said, growing hot at being challenged. "If we wanted to take it off the page, we would have taken it off the page. We didn't want to."

Deutsche Bank analyst John Inch, fresh off downgrading GE shares, was sitting front and center during the presentation and was dumbfounded by what he was seeing. "There was a stunned reaction," he said. "Jeff had clearly lost his composure."

The questions didn't get better. Was the Alstom deal not working? Could the Power division improve its cash flow? Would the company consider spinning off the Healthcare division, which Immelt had once run?

Immelt, as he had done before, argued that investors had GE all wrong and were mispricing a stock that should have been above $30 a share. The Aviation business was booming, outpacing competitors with its newest model. The once-troubled Healthcare unit was on the upswing. The oil and gas business, which had suffered through sliding crude prices, was now riding a rebound.

"It's not crap. It's pretty good, really," Immelt said of his company's financial performance.

When the grilling was over, Immelt wasted no time getting out of Sarasota. In less than an hour, he was aboard a GE jet. Immelt, his credibility on Wall Street battered as much by his besieged tone as his shaky assurances, limped through the rest of the week as he spoke privately with frustrated investors who were seeking clarity on the state of the company.

Ed Garden had recently made a rare public speech and declared that GE could actually exceed the 2018 goal Immelt had just waffled on. Undermined by his own CEO, Garden fumed. Among the calls from investors was one from Trian. The once-friendly activist made clear it would now push for a seat on GE's board.

Suddenly, one question that Immelt had batted away with little more than a joke during the questioning in Florida now seemed significant.

JPMorgan analyst Steve Tusa, who had been telling investors to sell GE shares, said to Immelt, "Hate to put you on the spot, but I'd like to get any update on succession planning, potential time," Tusa said. These were delicate matters that analysts didn't usually raise so publicly; it was telegraphed by GE staffers that Immelt loathed being asked about his retirement. But blood was in the water, as Tusa's jocular tone showed. "I know," he joked, "you just can't bear the thought of not coming down to Sarasota."

Immelt's unraveling on the stage exposed an inability to see that his leadership had limits. The near-death experience in the financial crisis created a pool of potential investors that wouldn't trust GE with their money while the same management team remained at the top. For them, the first step in any solution to GE's problems was that Jeff had to go.

48

THE NEW GUY

AFTER THE SARASOTA debacle, Jeff Immelt realized that he had lost the confidence of investors, especially Trian. Without that, even the eternal optimist saw little chance that he could stick around to lead a turnaround at GE. He wasn't going to get to the promised land.

For many investors, GE had lost its mojo long before. Its lackluster performance, fuzzy financials, and unknown risk just didn't fit with a lot of investment portfolios. Investors had owned GE for everything it wasn't delivering now — because they thought it was dependable and churned out quality products and quality people.

But after the crisis and Immelt's endless deal-making, a substantial number of investors weren't willing to take a chance anymore. They would, and had, put their money somewhere else. And they certainly weren't going to get back into GE with the same people running it.

Immelt found that concept hard to understand. His vision, his plan, his pivot — they were still what GE needed, he believed. In the eyes of many, the company really wouldn't change until he was gone from the helm. After the meltdown at EPG, he was finally realizing that it was time for that change, but he wanted to do it without being pushed.

The contest to replace Jack Welch had been a nightmare, and Immelt didn't want to replicate that long, drawn-out public battle. He was open about his desire for a less public contest, and determined to be true to his word.

The plan so far had worked. The board had long been watching the

final candidates closely, visiting with them, having them present to the board, and making sure they were in visible, consequential roles in the company. But even the candidates weren't sure that they were on the short list.

One problem avoided with a total secrecy approach was the aftermath of the announcement. Although top people and others kept out of the loop were likely to leave, the company didn't publicize the short list of candidates out of fear that it would be obvious that the succession process was coming to an end. Under Welch, the final candidates had become known to the public largely because he appointed their successors to work alongside them as chief operating officers.

Years earlier, GE's board had quietly set a target of late 2017 for a new CEO to take over, and it had identified four GE men as possible successors: Jeff Bornstein, Steve Bolze, John Flannery, and Lorenzo Simonelli, the lower-profile boss of the oil and gas business.

The candidates got a wake-up call when the *Wall Street Journal* reported in February 2017 that the four were leading the race.

A few months later, just prior to Immelt's disastrous performance in Sarasota, the board called the candidates to New York to audition. In a hotel in downtown Manhattan, the candidates made proposals on where they would take GE if they had the top job.

The organizers chose the location because they thought it was a site where no one would notice the GE directors. That strategy reflected the sense of self-importance that the company brought to the process. In reality, however, few people anywhere in the world would ever have recognized any GE board member other than Immelt.

By that time, the secret race had already been won. While the candidates were allowed to audition, Flannery was the unofficial heir apparent.

Much to his surprise and disappointment, Bolze had been ruled out long before, and he was expected to leave the company. His team, which had stretched so far in hopes of riding his coattails, wouldn't get the chance. Not only was his unit the increasingly sclerotic heart of GE's struggles, but Bolze, who had occasionally clashed with Immelt, was seen early in the process as a poor fit as CEO.

Simonelli, seen at forty-five as too young for the main job, was ticketed to run the public company that resulted from the merger of GE's

oil and gas unit and the oil-field service company Baker Hughes. He was about the same age as Immelt and Welch when they took the top job.

Because Bornstein hadn't run a GE business unit before, Immelt and the board felt that, if he would agree to stay on, he would be better placed as a partner to a successful candidate. Once a Wall Street favorite, some large investors were frustrated that Bornstein came into the job promising big changes that never fully materialized.

The decision to accelerate the succession process by several months was Immelt's decision. Since he had lost credibility and Trian was coming onto his board, he saw no reason to wait. The board agreed, especially knowing that Bornstein would serve as a partner as Flannery got his sea legs.

On a Friday in early June, the board gathered again to take a final vote. Secrecy remained paramount: the room had been swept for listening devices the day before, and security guards were stationed there through the night.

Instead of working in the GE office, a small staff gathered at the Boston apartment of human resources chief Susan Peters to write the press release and other materials for the announcements. Fearing that others in the office would see the commotion and make a guess about the news, the board was trying to avoid a leak to the media.

The vote was unanimous for Flannery, who got the call that afternoon. The announcement would come early the following Monday. The communications staff spent the rest of the weekend terrified that a call would come, especially from the *Wall Street Journal*'s Joann Lublin, whose connections to corporate boards were unmatched. She had deep roots from her nearly half a century at the paper. GE was so close to the finish line, and they feared Immelt's wrath if the news got out.

The news came as a surprise to many who saw Flannery as suitable for the role but lacking thorough experience in operating companies or divisions. He was running the Healthcare division, and doing well, but speculation had usually drifted back to Bolze and Bornstein. Immelt supported the decision to promote Flannery.

Flannery's experiences in those early days of being named CEO were similar to Immelt's when he started—a whirlwind of plane flights, handshaking, media preparation, and conference calls. But the soft-spo-

ken and analytical Flannery, more banker than salesman, was nothing like the affable Immelt; there was no trace in him of the swagger, booming presence, or charisma that characterized the outgoing CEO.

For Trian, Flannery was ideal, and a balm for their frustrations with Immelt. The new CEO had an investor's mind-set, crunched numbers naturally, and was obsessed with the cash produced by GE businesses. And unlike Bornstein, who would stay on as CFO, Flannery had no history of tension with Trian.

Flannery was a product of GE Capital, but he loved the company in its entirety. GE had allowed him to thrive and to carve his own way. Now he was at the top. He had spent years imagining a more streamlined GE and was bewildered by its inability to meet cost-cutting targets. Now, in his own understated but uncompromising way, he could make that happen.

The GE board had a view similar to Trian's and considered Flannery just what GE needed. They didn't feel that Immelt's time at the helm had been a disaster, but also increasingly felt the company needed a cold dose of candor and self-examination.

Flannery was well aware of Immelt's flaws, and he wanted to change the GE culture to encourage debate and focus. He didn't want to be an autocrat. Flannery had a lifer's faith in the value that was locked up in GE's vastest resource: thousands upon thousands of people. If he could tap them, the company would shine.

Some of Immelt's signature endeavors and buzzwords evaporated when Flannery ascended. It was tempting to cast Flannery as the anti-Immelt, but he too was well steeped in the GE culture. He had come from Capital, whose workers had felt marked as misfits in the company since the financial crisis ten years earlier, despite operating what its employees felt was a world-class finance organization that had outmaneuvered Wall Street rivals for years. Flannery was also deeply instrumental in the Alstom deal, having argued that it would be a valuable asset, despite his attempts to distance himself from it when he became CEO. John Flannery was being charged with taking GE in a new direction, but he was also deeply rooted in where the company had been over the course of the Immelt years.

In his first words after the announcement of his ascension, Flannery

made it clear that he planned a thorough review of the company — a typical move of any incoming CEO.

"I owe myself and the team and the company and you a couple of times around the track looking at the business in the coming weeks and months here," he said.

"I firmly believe our best days are ahead of us."

BEHIND THE CURTAIN

IT HAD TAKEN two corporate jets to take Jeff Immelt around the world. For much of his career, the famously globetrotting chief executive often had an empty business jet follow his GE-owned Bombardier or Gulfstream to far-flung destinations, just in case there was a mechanical issue that could lead to delays.

The spare planes started flying early in Immelt's tenure and continued to follow him in his travels into the months before his resignation. It was an unprecedented and highly unusual practice. Even heads of state didn't get such treatment, not because they were less important, but because it just wasn't practical. In case of emergency, there would always be other jets available on the ground.

This GE practice was kept quiet, and not even the board knew about it; the board only approved the budget and didn't get an itemized description of the corporate jet operations. It wasn't a secret that Immelt traveled constantly, so it had never been surprising that it cost millions to ferry him around.

Immelt initially denied knowledge of this two-plane operation, but then pivoted to say that he found out about it in 2014, when the practice stopped. But the practice didn't stop. Flight records revealed the well-worn pattern of the two jets often following the same path with itineraries only minutes apart. The secrecy betrayed the notion that it was a practical decision.

At small airports that catered to private jets, like those in Westchester

and the Hudson Valley, members of GE flight operations knew to keep quiet about the planes. The airports didn't have terminals but rather little buildings that had some seats and a restroom, with maybe a TV playing. During layovers and maintenance checks, when flight crews could easily mingle, GE crews had specific orders not to use certain terminology and not to talk openly about the two planes. They were also directed not to leave the planes too close to each other on the tarmac.

The extra plane's flight manifest sometimes listed "Robert Jeffries" or "Jeffrey Roberts" as the passenger on the second plane. A member of the flight team once earned a reprimand from superiors by openly referring to the extra plane as a "shadow plane."

Despite the confusion and finger-pointing that set in when the practice became public, the message had been clear for years: it would be detrimental to everyone at GE if the practice of the CEO traveling with two planes became public. Some warned that the practice was wasteful and reckless. According to one person, CFO Keith Sherin confronted Immelt about the practice. The CEO brushed him off.

Early in Immelt's tenure, GE flew spare jets when he traveled around the country, but it later stopped doing so on the CEO's domestic flights. On some trips, however, both overseas and domestic, the company would pay for charter jets to stand by at different destinations.

There were close calls. In September 2010, an anonymous Montana political blogger wrote that two GE jets flew into Butte, Montana, delivering Immelt to an economic summit hosted by then-senator Max Baucus. The blogger saw no apparent reason for the multiple jets, given the small GE delegation present at the meeting. He talked to airport staff, who said that the second jet was an extra, empty plane. Upon inquiring at the company, the blogger was told by a GE spokesman that the board required the CEO to use corporate jets for security reasons and that the claim of "some sort of 'chase plane' scenario is wrong." The blogger's report didn't gain any wider traction.

A few years later, the two-planes issue came up again, this time in a whistleblower letter to the board. Although the complaint was addressed to the entire group, the board never received it. Instead, the complaint was reviewed by an executive committee that included human resources head Susan Peters, Jeff Bornstein, and general counsel Brackett Denniston. The committee recommended changes, including using locally

chartered planes as backups instead of flying two GE-owned planes. The committee reported its findings to independent director Sandy Warner.

But now, as Flannery took the reins, this practice had become a big problem. Shortly after Immelt stepped down as chairman, the *Wall Street Journal* published an article about his practice of traveling with two planes. GE and Immelt scrambled to offer evolving stories about the details of the extra jet, leaking their explanation to press outlets with claims that the *Journal*'s coverage wasn't accurate or fair. In the end, they insisted that the practice had stopped in 2014. But again, flight records and other corroboration showed that it had continued until just months before Immelt left, as the *Journal* detailed in a subsequent story.

The rest of the board didn't learn of the practice until the *Journal* report in October 2017. Shocked by the news, the board launched an internal investigation, but Immelt stuck to his shifting story. The practice had been set up by the corporate air team, and he insisted that he had barely even talked to the managers of that group during his time as CEO.

The facts still refused to line up with the explanation. People who were closely involved in the CEO's travel said that Immelt was very particular when it came to the planes, with clear preferences about which plane he would use. He was intimately aware of how the organization worked. In the negotiations with officials in Boston over GE moving its headquarters to that city, maintaining quick access to the GE jets was a priority for Immelt.

Data showed that the double-jet trips continued until just months before the practice was exposed. A few months earlier, in March 2017, two GE-owned Bombardier Global Express jets took off from Boston within nineteen minutes of each other and flew to Anchorage, Alaska, according to government flight records reviewed by the *Journal*. One plane stayed in Anchorage for more than five days, while the other flew on to South Korea and China. Immelt made a very public visit to a Chinese factory during the trip. His plane returned to Anchorage on March 17, and within ninety minutes of his arrival, the records showed, both planes left Alaska to return to the East Coast.

The practice was also in place during times when GE was under investor pressure to cut spending and boost profits. Experts told the *Journal* that the extra plane added about $250,000 to the cost of a round-the-

world trip. The large jets are the size of a regional airliner and typically configured to seat ten to fourteen passengers.

GE's board did require the CEO to fly on the company's aircraft for both business and personal use, for security reasons. Many large companies have similar policies. And while the use of a private jet, or two, may seem unnecessary, it can be very useful for an executive who needs to get somewhere in a hurry and not waste time. Walmart founder Sam Walton was famous for flying his own plane to get around to the company's stores.

In the letter he ultimately wrote to GE's lead independent director, Jack Brennan, Immelt touted his success at the company and denied any awareness that he was accompanied by another plane in his travels.

For John Flannery, the two-planes controversy reflected what he already knew: GE wasn't the company that most people imagined. Presiding over the sprawling operations as CEO was a mind-bogglingly complex job, and it was easy to get lost in the details.

He turned to a network of former GE executives and asked them to talk about their views of the company. Regardless of their opinion of Immelt, everyone was happy to help Flannery get the best shot at running GE. Flannery even went to see Jack Welch on Nantucket. Some expected Flannery to be more like Welch and less like Immelt. In the Aviation division, some workers responded to the news of Immelt's replacement by chanting "Jack is back." The enthusiasm was double-edged — an endorsement of Flannery and a rebuke of Immelt.

Before taking over, Flannery traveled around the company to review operations and meet people, following a playbook he had used when taking over the Healthcare business. An inventory of the operations, he believed, would allow him to reorganize in the most efficient way possible. The problem was that, with what he was discovering, he had less time to do that than he had thought.

He still wasn't sure if Bornstein was aware of the problems at GE. The CFO was insisting that it was all news to him, but it was hard for Flannery and some others to believe that an imperial CFO like Bornstein, with his matrix of underbosses, hadn't known what was happening. Given that Flannery needed to trust his partner, however, and given what he had seen of the company, he thought that it might not be that

far-fetched that misbehavior had been taking place right under everyone's noses.

The lack of cash was disconcerting enough. Aside from sizing up the divisions, Flannery spent the weeks leading up to his official start on August 1 meeting with dozens of major investors, and he heard their concerns too. For many, Flannery's financial mind-set was comforting. Unconvincing company buzzwords, like "Predix" and "the GE Store," had simply dropped out of these conversations. When investors said that they wanted Flannery to move quickly, he told them he would need four months.

The issue was that the structure of GE made it difficult to see every aspect of its businesses at once, a complaint dating back to the days of Welch. The company had also made a habit of rearranging its business units every eighteen to twenty-four months, making comparisons and deep analysis painfully difficult. As he tried to understand the other businesses that were part of this siloed structure, Flannery found that his three decades inside the company, even running one of the major industrial divisions, were not enough to make the workings of the whole intuitive — or its biggest problems obvious. Another reshuffling of the portfolio was under way — Power merging with the Energy Connections unit — meaning that the complexity and opacity would only increase as demand for GE equipment kept falling.

On top of that, Russell Stokes, the highly regarded executive who replaced Bolze atop Power, was also running the Energy Connections business and now would be handling both divisions. Stokes was given an impossible hand to play, and he essentially had a weekend to adjust to the overloaded mission.

Stokes, who had spent more than two decades at the company and previously ran the transportation division, grew up in Cleveland going to see GE's Christmas lights at Nela Park. Like Flannery, he scowled at the typical flood of PowerPoint presentations that came with working at GE. He would tell people that a long presentation was a bad sign, usually indicating that the presenter was hiding something. A well-run business didn't need a lot of slides, he said, just the facts.

Flannery hadn't picked Stokes for the job, but he was now depending on him to help turn around Power. The two of them had seen eye-to-eye on the condition of the business. Bolze and his team had let inventory

pile up, hoping the market would turn around and leave GE in a prime position to sell scores of turbines. But the power market was going in the other direction, and GE now had a lot of cash tied up in that inventory. The biggest problem was that a miscalculation on inventory couldn't be easily or quickly reversed. GE, like its customers and its suppliers, planned its manufacturing out months in advance. Even now, long after the folly of Bolze's investments was clear, it would take even longer to unwind the massive oversupply of turbines and parts that Power had accumulated.

Power was the biggest division, and especially in the wake of the Alstom deal, it simply had to succeed for GE to be healthy. The other divisions could help offset the issues, Flannery knew, but that left little room for error. And he sensed that slim margin was about to disappear.

Toward the end of July, GE reported its second-quarter financial results. The conference call with investors was awkward. Immelt was on the line along with Flannery and Bornstein. The executives were clear that the Power division faced shifting winds, demand would be down for the year, and it now expected a decline for 2018 too.

Immelt, with just ten days left as CEO, pushed back with his customary optimism. "We have pressure in Power and Oil & Gas, but we will outperform on structural cost-out," he said, using the GE lingo for cutting fixed costs in a business. He mentioned the same idea later, only to be followed by Bornstein nudging back.

"There is both an opportunity and a necessity to restructure the cost structure of this business, given the market that we are operating in," Bornstein said. The executives went back and forth as Immelt injected his optimistic view and others pushed back with a more measured view. Flannery noted that he would present his assessment in November.

In what seemed like a parting shot in his last appearance to investors, Immelt professed dedication to the dividend. He reminded investors that the worst day of his life had been the day he cut the dividend in the crisis. He said that GE prioritized the dividend, "whoever our CEO is." Meanwhile, Flannery was coming to the exact opposite conclusion: GE wasn't making enough cash to cover its dividend payments and hadn't covered it in years. Immelt's pivot wasn't working. And GE couldn't afford to keep shoveling billions out the door each quarter to investors, as much as they had come to depend on their dividend payouts.

Perhaps the most ominous moment of the earnings circus was what happened prior to the news. As the executives were preparing and rehearsing for the event, Jeff Bornstein spoke up to the investment relations team.

"Oh, we should mention the insurance review," he said.

Flannery's head immediately turned — insurance? GE had said that it sold all its insurance holdings long ago.

50

TRIAGE

Like a politician winning a first election, Flannery had become recognizable. After sixteen years of Immelt, Flannery was now considered GE royalty.

He hadn't exactly been a peasant before. Flannery had been a rising executive at the company for years; he'd overseen GE's operations in entire countries, run high-profile deals, and known well the pressure and comforts that came with GE's highest offices. He might not have been as enticed by the formalities as some others, but he was getting more comfortable being whisked from one place to another, having a security detail, and constantly having a car, plane, or helicopter waiting to deliver him to the next stop.

Flannery could work a room like Immelt, but their styles were enormously different. Flannery visibly relaxed and brightened when he spoke about what engaged his personal side — a book, a news event, or family. Interjections, whether serious inquiries or social introductions, made him visibly tense back up to what seemed a naturally guarded state. The superficial, even in business settings, didn't hold him.

Immelt was headed for the Small-Talk Hall of Fame. He wielded a masterful recognition and ability to bring a room to a low chuckle. Flowing through groups in a crowded hall, like a good politician he left everyone feeling he had talked directly with them. Immelt's allies said that the trappings of GE royalty meant little to him. And yet the very

environment he moved through for more than a decade changed to his tastes without his asking.

Immelt notoriously preferred cold temperatures anywhere he had to present to an audience. It was a subject of humor inside and outside GE, but also not a joke. No effort was spared by the staff to ensure that meeting venues were cooled to meat-locker temperatures to accommodate the boss's preference, irrespective of whether anyone had ever heard him make such a demand out loud.

Was a CEO supposed to object that the temperature was not to his liking, or demand that elevators were always open and waiting for him? Or that the cold diet sodas he liked were always present on a sideboard when he entered a room, no matter how far-flung the visit or conference room he walked into? Some of his colleagues and underlings noticed the preparations, and even without hearing them demanded, the lesson was the same.

The perks weren't why anyone did the job. But the environment of supplication that surrounded the head of the company was also nearly impossible to opt out of. The room must be made ready for the boss.

Immelt made no secret of his distaste for waiting. Interruptions in a meeting from a ringing cell phone produced serious scorn: "What *is* this?" he would growl at the perpetrator with a sharp glance that conveyed a stinging disapproval. When flying the company helicopter into Crotonville, he would often have a black car waiting to take him from the helipad to the front door — a distance of several hundred feet.

But now, Jeff Immelt was gone. He was still chairman, but he had essentially disappeared from the company. He was somewhat in the way, and he could sense that. There was little for him to do, and his office was moved from corporate headquarters and into a WeWork space that was being used by an offshoot for the GE Lighting business across town in Boston.

Before officially stepping down as CEO, Immelt was already trying out for another job. He was in San Francisco being considered for the top job at Uber, the ride-sharing service that was in the midst of a governance crisis and power struggle with its founder. Immelt presented himself as someone who could be "the grown-up in the room."

Media attention to the Uber selection process was unending, including reporters staking out Immelt at his hotel. Eventually realizing that the job wasn't going to be a good fit — and also knowing that he didn't have the support he needed to get the job — he withdrew himself from consideration to save face.

Back at GE, Flannery hit the ground running and made big changes within a few days of officially wearing the crown. Consciously or not, a lot of moves cut into the fat that had been gained in the recent decades of the company's growth but not revisited as the company evolved.

As one early cost-trimming move, Flannery grounded the corporate jets and put them up for sale. He directed employees to use charter flights or just fly commercial. He also eliminated the company policy of providing a car to its top eight hundred people.

He delayed parts of the Boston headquarters project, sending shivers down the spine of local officials who had bent over backward to bring GE to town. GE insisted that it remained committed to Boston.

Flannery canceled the lavish annual three-day, invitation-only GE retreat at the Boca Raton Resort & Club, a networking event for GE leaders gathered from around the globe to spend afternoons in Florida golfing and fishing. Hordes of GE's top brass had once gathered in Boca, many in the Immelt-era company uniform: loafers and Dockers, crew-neck sweaters over button-down shirts. Some described the meeting as right out of *30 Rock*, a sitcom based on NBC and its former corporate parent GE. On the final evening at Boca Raton, the CEO would dole out sought-after internal awards.

Instead of gathering in sunny Florida, Flannery replaced the event with a meeting in Boston in January. There would be fewer attendees and certainly no golf.

He began the shutdown of research centers in Shanghai, Munich, and Rio de Janeiro, shifting some of their engineering work into individual business units. The once-sprawling and endlessly growing research operations would now be concentrated in just two global research sites, located in Niskayuna, New York, and Bangalore, India.

A picture of the company started to come together. The operations weren't in the shape they needed to be. Few had doubted the decision to jettison GE Capital, but the plan to replace its earnings had fallen flat. Industrial growth wasn't coming fast enough, and the company had

blown billions on share repurchases that now seemed worthless. During Immelt's tenure, GE spent well over $100 billion buying up its own stock, much of it at a price far above where it traded now. Without adjusting its sacred and expensive dividend, the company had gutted its ability to generate cash.

In the first six months of 2017, GE had earned hardly any of the $12 billion in cash it had previously said it would take in that year. It would need at least $8 billion just to cover the dividends it had promised stockholders, before getting to the myriad other areas, like research and development, where regular infusions of capital were essential to its success. The path was becoming clear — but heading to a place that had been inconceivable just weeks earlier.

CLEANING HOUSE

IN AUGUST 2017, the top GE executives descended on Crotonville for their annual leadership gathering.

Many visited the campus at various times during the year, taking classes and teaching classes. There they mingled with middle managers, customers, and executives from other companies who wanted to learn the GE leadership magic. The summer leadership meetings usually left executives refreshed and reassured that the foundation of GE's success was not merely power turbines or jet engines so much as the people in that room — the managers groomed in Crotonville who believed that they could enter any industry, anywhere, and dominate it.

In 2017, the annual pilgrimage was different. The stock price had been slumping, and many in the company were still trying to get a picture of the future of GE in the wake of Immelt's sixteen years at the helm. Flannery, they knew, had started a months-long review of every corner of the company, and the company's uncertain state had made a lot of people nervous.

On that summer afternoon, the auditorium buzzed with whispers about what lay ahead. Regardless of the uncertainty, no one doubted the 125-year-old company's ability to rise again. It always had.

Flannery addressed the group, laying out many details of the company's operations that he was finding out for the first time. In stark and jarring contrast to Immelt's shining positivity and predictable pep talks,

Flannery said that the state of the company was bad and the future was probably worse. Immelt had never given a speech that portrayed the company as facing danger, or even dire challenges. But it wasn't Flannery and his sobering comments that the attendees would remember later about that day.

Next came Jeff Bornstein, who many had once thought would be CEO someday. He had been a counterbalance to Immelt and was generally respected by Wall Street. Many in the room felt comforted that he had committed to GE and to helping Flannery navigate the job.

Now Bornstein launched into an exhortation: Run the company like you own it. Be the leaders General Electric bred you to be. You should all be accountable for every prediction made and every target missed.

"I love this company," he said. Then he stopped and took a breath, deep and halting. He started up, and stopped again. Jeff Bornstein, the muscled, nicotine-gum-chomping, weightlifting, hard-driving CFO, was . . . crying.

The reaction in the room was confused and inconsistent. Some thanked Bornstein for his passion, while others left the bizarre scene wary. If this guy was fighting back tears, something must be seriously wrong. The group shuffled out in shock and confusion. Flannery and Bornstein had hammered away at some of GE's intractable bad habits, but there had been no bombshells revealed. Still, stress and strain had reduced one of their toughest tough guys to tears. The moment would linger in the minds of GE people for years — a revelation, in retrospect. That was the moment many of them realized GE was taking on water, and that no one knew how bad it was going to get.

Investors by now were braced for the company to cut profit forecasts and possibly reduce its dividend, despite promises by GE that the payout was safe. For some, shrinking the dividend would have come as a positive move because it would alleviate cash fears and signal that management was taking aggressive action at last.

Flannery agreed, but thought he needed more time. He simply didn't have all the information he needed.

If Immelt was known for his vaulting optimism, Flannery soon became known for his indecision and endless analysis. Few decisions, even

major ones, were final. A critical strategic move, like the separation of a major division, could be made, only to be reassessed at any time. Flannery's style was quickly grating on top executives who worked with him.

The complexity of GE and its problems required true commitment to decisions; otherwise the process would get bogged down. Searching for flaws in his own reasoning, Flannery regularly sought input from outsiders. But an abundance of feedback prevented any path forward from looking crystal clear, and the result was more indecision. Flannery was always conferring with the board and openly encouraging debate. Some directors grew frustrated.

Those on the board who worried about Flannery and his lack of experience running the company found comfort in the slate of advisers who were sticking around. Company veterans John Rice and Beth Comstock, both vice chairs, could lend a guiding hand, and Jeff Bornstein, the freshly minted vice chair, served essentially as a partner to Flannery. But as summer turned to fall in 2017, that network was quickly purged.

On the outside looking in, Immelt had chosen not to stick around for the new CEO to dismantle what he had left behind rather than risk a standoff with Trian. Immelt stepped down as chairman in October, months earlier than expected. He broke, at last, with the company and Flannery, and the board.

Flannery started making big changes at the top of the company. He started with those advisers who were high-ranking veterans of the Immelt era. Comstock and Rice were shown the door.

Comstock would later say that Flannery personally fired her and that she was shocked by that. But Comstock had already told people that she had long wanted to leave and stayed only at Immelt's request. Beth was distancing herself. And Flannery wanted to run the business without worrying about fitting the strategy into a story line or selling that story to the public. He was closing the GE Store.

Even after Flannery said goodbye to Immelt and fired his holdovers, his third move was the most stunning. It became public in a late Friday regulatory filing to the government.

As the board was gathering in October for a monthly meeting, Flannery stepped into the room to make an announcement: Bornstein was resigning. Bornstein himself later came in to explain his decision. GE would probably have to offer Trian a seat on the board. His departure

before that happened might spare directors some conflict between Trian and management. Bornstein would leave GE along with Comstock and Rice.

This resignation blindsided several directors, who were disappointed that the board hadn't been consulted. They felt that they could have persuaded Bornstein to stay on. Investors were also worried by the CFO's resignation. In previous corporate meltdowns, the sudden resignation of the CFO was usually a bad sign that hinted at deep problems.

The next big move came on the following Monday: GE named Trian's Ed Garden to its board, giving the activist a direct say in decisions and access to its financial details. The move, stemming from GE's failure to hit the agreed cost-cutting and performance targets, was the same deal hatched in Bornstein's Back Bay townhouse months before. GE's failure to hit its targets had finally let a wolf in the door.

Flannery and the board wanted to avoid the distraction of a potential proxy fight, and no directors opposed Garden's selection. Some welcomed the new voice into the discussion, even if Garden could be abrasive at times, while others on the board bluntly declared their distaste for the activist. He was fond of telling the group that Trian had lost $1 billion on their watch. James Tisch, the longtime director who ran Loews Corporation, bought $54 million in stock partly to rebut Garden's exhortations that the board didn't have skin in the game.

Now on the inside, Garden was in a familiar role for him and his Trian colleagues. The GE investment was perhaps the only time that the fund had invested in a business without obtaining a seat on the board of directors. In most of Trian's investments, they played the role of "highly engaged shareowners." Garden and Peltz liked to say that they brought a private equity mind-set to the public markets — reviewing numbers alongside management, second-guessing decisions, and setting strategy.

Immelt and some of the more entrenched directors scoffed at letting Trian on the board. Once Garden took his board seat, there were open arguments and plenty of tension. In his first board meeting, he dressed down a junior executive who was giving a presentation. Some of the directors respected his intelligence and experience and valued his opinion, but seethed at his communication methods. They felt that he talked down to colleagues, and his view of them as ineffective was insulting. It also was becoming clear that a lot of directors would lose their seats.

Other directors warmly welcomed Garden, even if he was a complicated character. He could be blunt and gruff, but he was there to help solve the numerous problems facing the company. Trian had already lost a lot of money trusting GE leadership. These directors recognized that pushing the board in new directions was Garden's job. As the company collapsed and the stock dropped, Garden personified the frustration and anger directed at the company by all investors.

With some directors sleeping during board meetings and some seeming more interested in the corporate jet and other perks, many on the board recognized that it had become dysfunctional and were eager for Trian's presence. Wall Street agreed. "Nelson Peltz is a genius," tweeted Josh Brown, the financial adviser and investment pundit. "When he asks for a board seat, you give it to him and say thank you for helping."

In Immelt's absence and with new challenges looming, some directors were ready to move on. For example, Lowell McAdam, the sixty-one-year-old chairman and CEO of Verizon, was having concerns about remaining on the GE board. McAdam had run a tight ship at Verizon and shown an unceasing ability to find cost savings in the operations. He was a Cornell-educated engineer and had served six years in the US Navy. McAdam had taken a GE board seat as he began to think about stepping back from Verizon and his years of nonstop work. And a GE directorship was still considered an enviable sinecure. McAdam had been honored and excited to join the conglomerate. As a car collector and general lover of machines, he was intrigued by GE's heavy industry.

But the experiences of his short stint sent him running. The board was openly dysfunctional, problems at the company were deep and troubling, and serving GE in this hectic period was more time-consuming than he had bargained for. The GE board typically huddled a dozen times a year. Under Flannery, it convened nearly fifty times.

When it became clear that Flannery was going to shrink the board, McAdam saw his opportunity to step down. He had a company to run. He wasn't sure what was going to happen to GE.

Garden, meanwhile, wasn't coming onto the board alone. This was Trian's biggest investment ever, and it had an entire staff devoted to the GE operation, including an analyst who essentially spent all his time on GE.

They ripped every layer of the onion apart once they were inside,

with few limits. Being on the board gave Trian substantially more insight into the company, but it also presented limitations: now considered insiders under securities law, Trian would be prohibited from acting on any nonpublic information it obtained by being on the board. Ed Garden had the raw numbers he needed about GE's inner workings. But now his hands were tied.

52

CONSERVATIVELY CALCULATED

THE PACE OF change felt hectic inside GE, but outside its walls the absence of any signs of action was driving away investors. By the time the third-quarter results came in October, the stock was below $25, and it was quickly losing ground. GE warned that full-year cash flow from its industrial businesses would come in at just $7 billion, a massive miss of the previous projection of $12 billion.

The loss stemmed almost entirely from GE Power. As the service business had tweaked contracts to make them more profitable, it was boosting revenue and earnings but also delaying the inflow of cash. Meanwhile, the business was misreading the market for new turbines, building up expensive inventory, and further depleting its cash reserves.

The results were discouraging, and its plan for recovery was unclear. Too often, Flannery felt, the company under Immelt had failed to be adequately rigorous in making major decisions about how to spend its cash. And now, because of the opacity of GE's decentralized structure, Flannery felt he needed more time for analysis and scrutiny of the disparate business units.

After working for years for a CEO who disliked dissent, some top GE executives found Flannery's candid approach invigorating, but others quickly became unsettled and annoyed. Even Jamie Miller, the new CFO who had previously run the transportation division, would eventually grow weary of Flannery's cautious approach. She was part of a group

that thought he was being too negative and felt that his overcorrection for Immelt was making it harder to solve GE's problems.

Regardless, Flannery was going to do it his way. His stubborn, deliberative approach stemmed in part from the belief that he had time on his side. CEO of GE was a long-term appointment, and Flannery felt his duty required him to make absolutely sure he was taking the correct course.

He preached that there were "no sacred cows" in his strategic review of the company portfolio. He had made a similar vow several years earlier when he led the deal-making team that moved to shrink GE Capital and sell GE Appliances. For Flannery, there was no room for sentimentality in operating a business, especially one under duress.

When the heads of GE's major businesses presented their 2018 budget plans and projections, Flannery sent some of them back to redo their numbers and make even deeper cuts. At headquarters, he was clear that GE would be more open about its problems and that his turnaround efforts would be best measured by the company's share price.

Mid-November came fast, and the investor update loomed for the new management team. All of Wall Street and the business press expected a major new plan for GE from the new CEO. Even though some advisers thought it was too early and worried that a final plan hadn't solidified, Flannery, under unrelentingly heavy pressure from investors, was forced by the scheduled event to set a course.

Hundreds of investors, analysts, and reporters converged on a nondescript building in Midtown Manhattan and gathered in a wood-paneled room. The tone had been set hours earlier, as the sun rose, by a massive disclosure: GE would cut its dividend in half. Now it was time for details.

Flannery put some blame on the previous management of the Power business and alluded to Immelt's shortcomings by declaring the failure of the so-called pivot strategy: "We've been paying a dividend in excess of our free cash flow for a number of years now," he told the investors and analysts. This was a judgment that sounded even worse in plain speech. The supposed excess earnings the company had doled out to its shareholders had actually been borrowed. It was a dividend in name only.

· · ·

Flannery's admission at the investor update meeting was totally shocking. GE had been sending cash out the door to repurchase its stock but wasn't bringing in enough cash from its regular operations to cover its dividend. GE had the finances to write those checks, but the practice wasn't sustainable. Buybacks and dividends are generally paid for with excess cash.

Large companies regularly repurchase their own stock. The practice began in 1982, thanks to changes in securities rules. Buybacks can be a way to shrink the total shares in the market, thus reducing dividend payments; they can also offset shares issued as employee compensation. Buybacks can also shrink share counts to help nudge per-share earnings higher and, at some companies, produce healthy executive compensation.

Repurchasing stock is also a major capital allocation decision. Every dollar's destination influences the success of the business. A dollar can be invested in the business, or spent on a deal, or used to repurchase stock.

Buybacks are controversial. Some see in them a lack of good ideas from management. Others see them as a kind of second dividend, a way to "return cash to shareholders." The approach to deciding whether to buy back stock advised by Warren Buffett and his longtime business partner Charlie Munger has two simple prongs: a company must have plenty of capital to run its operations, and the shares it repurchases must be selling "at a material discount to the company's intrinsic business value, conservatively calculated."

It's the second requirement that can be tricky. As the Berkshire Hathaway gurus note, "what is smart at one price is dumb at another." And they flag perhaps the biggest risk of all: "Many CEOs never stop believing their stock is cheap."

Buying back shares instead of making deals that significantly boost a company's earnings can have disastrous consequences (assuming, of course, you would otherwise have made good deals). Repurchases provide easy money, however, if the shares are undervalued, as Buffett and Munger preach, because you are essentially buying a dollar for less than full price. But that only works if you buy when the shares are cheap.

If the shares fall, buying back shares is little different than simply set-

ting the cash on fire. Paying a dollar for something that will be worth 80 cents in the future isn't a great way to spend someone else's money.

Jeff Immelt will not be remembered for wisely deciding how to spend GE's cash. Buybacks were a regular fixture under Immelt, who spent more than $108 billion on them after 2004, when the SEC required companies to disclose their practices. At the end of 2018, GE's entire market value was $67 billion. In Immelt's last eighteen months as CEO, he spent almost $26 billion in cash on repurchases even as the stock fought to stay near $30 during that period. Just fifteen months later, it had dropped below $10.

When Immelt's retirement was announced in June 2017 and the repurchases stopped, it was as though someone had grabbed the assailant's hand. GE spent $153 million in June and only $18 million in July.

53

A RESET YEAR

FLANNERY PAINTED A dire picture. It would take years to fix some of the company's businesses, with most of the work needed in Power, and he was going to cut the dividend in half, saving $4.2 billion a year. Immelt would not have made this move. Flannery also slashed Immelt's financial targets for the next year, finally euthanizing the delusions of $2 a share in earnings. The new boss cautioned that conditions in 2019 would also be difficult. He called 2018 "a reset year" for GE.

Flannery wanted to focus on three core divisions — Power, Aviation, and Healthcare — and exit most of the rest of GE's operations. As the *Journal* had reported weeks earlier, GE would sell its transportation division, one of the oldest and biggest makers of diesel locomotives. It also hoped to finally sell GE Lighting, as it had been trying to do for years without luck. Both divisions traced their roots all the way back to Thomas Edison.

Flannery was also shedding one of Immelt's biggest and most disastrous projects, the oil and gas business. The years of deals had been poorly timed, and the service side of the industry was now dominated by two giant companies, Schlumberger and Halliburton. It wasn't Immelt's fault that the entire oil sector had turned south. But he was responsible for GE investors being so openly exposed to the collapse. Just months after merging the oil and gas division with Baker Hughes into a

new company, GE would look to sell its two-thirds stake in the $40 billion company.

In a sop to investors, GE's board of directors would shrink from eighteen to twelve members and the existing board would be refreshed with new members. Jeff Immelt was gone, but his board had remained, and now there was little reason to maintain its structure.

These were the first steps in Flannery's plans to streamline the entirety of GE. Although he stopped short of breaking up the company, or even radically restructuring it, he planned to regularly review the structure of the portfolio and wasn't ruling out a breakup in the future.

Flannery's proposals were a sharp divergence from GE's usual tone of breezy superiority. His candor alone was a departure. Acknowledging the need for major structural change might have been the sign the market was waiting for that GE had learned and was starting to change. But that wasn't the lesson the market took.

Instead, GE shares fell below $20.

Flannery's candid description of the bleak state of the company and the needed restructuring moves suggested that the company was in even worse shape than anyone previously thought. As they speculated about the state of the company, notwithstanding what he had just proposed, analysts and talking heads wondered whether Flannery was doing enough.

Meanwhile, the continued uncertainty and the dropping stock price were crushing morale inside the company. The stock options held by employees were worthless, and Flannery seemed to be saying that GE would tread water for a while. The economy was doing well, but GE workers, especially those in Silicon Valley, were heading for the door.

Flannery insisted on transparency and honesty and tried to communicate that to his employees in his weekly iPhone-created video message, when he openly answered their questions. His tone and outlook on management were pragmatic. GE had the tools and would fix itself, he told employees, but it wouldn't be easy or enjoyable. Under Flannery, the message reverberating around the headquarters and the entire company was: no more success theater.

In other words, no more pretending that broken software was working.

No more making deals at any cost.
No more pretending that vague concepts and metrics had value.
No more sticking to the marketing story.
No more infallibility.
No more mythology.

54

A BILL COMES DUE

J EFF IMMELT KNEW John Flannery had been trying to reach him.
Since the succession, the two men hadn't talked much, perhaps only
a handful of times. Immelt had other things going on.

As bad news tumbled out of GE, Immelt had taken to joking that peo-
ple in Silicon Valley, where he was hanging out these days, didn't read
the *Wall Street Journal* or watch CNBC. He had good reason to ignore
the news. Many talking heads were pointing their fingers at the former
CEO, and the news coverage was doing the same in stories that revealed
the depths of dysfunction at the company.

Since stepping down as chairman, Immelt had been spending a lot
of time in California. He wasn't ready to retire yet. He was only sixty-
one years old and felt he still had something to offer. So he had made
a high-profile but hapless run for the job of CEO at Uber, the troubled
taxi startup. And he told all comers that he wanted to do something in
healthcare, or involving new technology. Immelt was looking for a sec-
ond act as the éminence grise to a startup, one that would give him a
chance to be part of medical, technical, or biological breakthroughs. He
gave the impression, to those who knew him, of a man interested in a
chance at redemption — another chance to prove his brilliance as a cor-
porate leader — even as he continued to reject the notion that failings at
GE had been his responsibility.

But the reason John Flannery was trying to reach him had nothing to
do with Immelt's future.

It was about GE's insurance holding — the holding that few even knew existed. Everyone — reporters, analysts, investors — thought that the company had sold the insurance businesses long ago, significantly derisking GE Capital. In often highlighting this point, Immelt and his top executives hadn't minced words: GE was out of insurance.

GE had indeed exited insurance — but was still exposed to large risks because it hadn't been able to get anyone else to take over the worst parts of its business. GE had spun off most of its insurance holdings into Genworth Financial in 2004 and sold most of the remainder to Swiss Reinsurance Company two years later. Still, there had been one part of the GE insurance book that the new leaders of the stand-alone insurance company weren't willing to take with them.

Immelt knew there was an ongoing review of the company's holdings in long-term care insurance, which covers costs like nursing homes and assisted living facilities. GE Capital had written long-term care policies until 2006, but doing so had proven to be a disaster for insurers because of the poor design of most of these policies and the drastically mistaken assumptions on which they were based.

The long-term care policies were a legacy business from the time when GE reinsured loads of such insurance policies (that is, they covered the payouts for the policies under certain circumstances). These badly structured policies usually became a nightmare for pretty much anyone selling them. Massive losses were recorded, and some companies were taken down by the coverage they sold.

As he tells it, Immelt asked for a review of the holding and the potential need to increase reserves as the portfolio showed that premiums were going to cover expected payouts. The preliminary estimate of the needed reserve fund, he said, was $2 billion to $3 billion. That was a lot of money, but not in Immelt's GE, with its persistent view that such small developments were almost irrelevant because of GE's size and ability to offset such pressures. Remember, $1 billion can be either a great deal of money or immaterial depending on whether it's a sales contract or a noncash adjustment.

In actuality, no one had looked closely at the policy details for years and no one knew what assumptions were used. When auditors took them apart and rebuilt models with more conservative assumptions, the

outcome was disastrous. GE needed $15 billion in cash to cover its liability.

The number was so large that it was hard to comprehend, and Flannery certainly couldn't make sense of it. It was the worst possible development for him. GE simply didn't have the money.

The company was stretched thin and still regularly ran close to $20 billion in commercial paper per quarter just to have enough cash to fund all its obligations. While GE could theoretically write a $15 billion check, it had to carefully manage its cash.

Flannery picked up the phone to call his predecessor, but never reached him. Immelt would have had little advice for Flannery and could only have repeated his polished assessment of the portfolio. Immelt would later privately admit that he would never have spent cash from the Capital exit on stock repurchases if he'd known such a liability existed.

GE had to answer to the Kansas insurance commissioner, an elected position. Because of the number of people insured, state regulators generally want to avoid the failure of an insurance company.

GE had constructed its insurance portfolio through deals that provided generous cash flow as younger policyholders paid their premiums. With that lucrative upfront cash from relatively young policyholders now causing an exponentially larger hole in the balance sheet, and with the cost of covering nursing homes and long-term care proving to be higher than expected, the worst was yet to come for GE.

The company began talking with officials in Kansas about finding a way for GE to meet its obligations without a huge blow to its cash holdings. The talks for relief started in December 2017 and didn't end until January 2018. By that time, GE had lined up $13 billion in credit lines to draw on, just in case the regulators insisted on full payment.

As it turned out, the safety net wasn't needed. GE won a waiver from regulators to allow it to build up the reserve over seven years rather than all at once. GE also agreed to conduct a more regular review of its portfolio.

But the news of the gigantic expense had traveled like a shock wave through the company. Several board members were appalled that this

could have ever happened. GE risk managers' most basic task should have been testing a worst-case scenario for its toxic assets and then putting the results on the radar. The potential risk should have emerged somewhere.

Directors also wondered why they had thought that the company was out of the insurance business. There were whispered questions about how something like this could slip through the cracks at GE. It just didn't make sense, and Flannery was among the surprised. Answers weren't to be found.

The core problem was that GE had made some bad decisions in reinsuring the long-term care policies. Once it was clear that the company had amassed a collection of some of the worst insurance assets possible, there was no chance it could find a buyer. A Justice Department investigation of the holdings was also making the asset impossible to sell. Meanwhile, as former employees described it, lax oversight and buried risks would keep the company from realizing the worsening results over the years and booking bigger reserves.

GE has denied allegations of fraud. As of early 2018, GE was reinsuring about 300,000 long-term care policies, about 4 percent of all such policies in the industry. Incoming premiums still weren't covering payouts.

"I wouldn't say what I saw was smoke," a former GE Capital employee told *Wall Street Journal* reporter David Benoit. "But there was enough concern that I decided to leave, because I didn't want to be there when there was smoke."

This individual had been hired to help improve the governance of the reserve process at GE's insurance business, but left after becoming concerned that senior executives in the division were changing numbers and their methodology without providing supporting evidence, the *Journal* reported. Another former GE Capital employee said that it was clear at the time of the 2004 Genworth IPO that a block of long-term care policies was toxic.

The bankers on the deal thought that the long-term care business couldn't be part of the Genworth spin-off. To make the deal more attractive, GE agreed to cover any losses. Genworth's regulatory filings made

it clear that GE retained the risk to the policies, which "do not meet our target return thresholds."

Genworth was eventually hit by its own long-term care problems to the tune of more than $2 billion. Other insurers also took charges for their long-term care policies. GE didn't make adjustments until Flannery was CEO.

The question of whether GE knew of the emerging problems has generated lawsuits as well as other questions. What is clear is that GE sold out of insurance but kept a small piece that many investors, executives, and directors weren't aware existed. GE has defended its actions.

In a company of GE's size, many developments and details are never disclosed to investors because they simply aren't significant enough to have a noticeable effect on the financial results. The company removed the disclosure of the long-term care liabilities from its annual report for 2012 and didn't put it back until five years later. In court filings, GE described the exclusion as a "normal business decision" and asserted that reversing the policy didn't mean that the previous financial statements were fraudulent.

Jeff Immelt and Keith Sherin had often described their successful exit from the insurance business. Immelt said that GE might not have survived the financial crisis if it hadn't shed the insurance operations, and he and his supporters used this example to demonstrate his astute deal timing.

Details suggesting a less than complete exit leaked out slowly at first. In early 2017, Jeff Bornstein started talking about a "stub insurance piece," which is jargon for the remaining holdings after substantially all of a business has been sold. He said that this piece had "virtually no returns associated with it." Bornstein's mention of this insurance was more serious that July, when he held the first call about it with Flannery, who only learned of it shortly before the call. By October, Bornstein was telling investors that Capital was suspending cash payments to GE.

Jamie Miller, who inherited the troubling insurance problem, mentioned the insurance risk deep inside the disappointing three-hour presentation in November, adding a little-noticed warning: the ghost of an insurance business that investors thought the company had rid itself of years before would prevent GE Capital from paying its dividend to

headquarters. The insurance hit, she noted, was likely to exceed the $3 billion dividend payment.

Two months after Miller flagged the $3 billion, it was clear that the problem was a great deal larger. GE was preparing for it to be more than $6 billion and needed to come up with $15 billion in reserves required by regulators to cover possible costs in the future. The figure was gigantic — almost four times the yearly cost of GE's new slimmer dividend.

ASLEEP AT THE SWITCH

N O ONE BLAMED Flannery for the insurance mess, but he was the one who had to steer the ship through the wreckage. Ever the realist, he painted a grim picture, his frustration obvious. Like Immelt, he thought he was making progress, but things didn't go quite as planned.

He described the insurance situation like a city engineer worried about a dam breaking. The damage was bad, but GE Capital was "buttressed" by the lucrative airplane leasing business. "We strongly believe this is contained inside of the GE Capital framework," he told investors.

But the debacle had evaporated GE's remaining credibility about its financial situation. The insurance problem prompted Flannery, who had declared in November that no stone had been left unturned, to reconsider his course. Like many longtime GE executives, his love for the organization ran deep, and he struggled with the prospect of being the one who would have to dismantle it.

At the advice of the board, he also signaled that the review of the entire GE business would continue, including a real consideration of breaking up the company. Subsequently, Flannery was upset that his statement was interpreted to mean GE might break up — partly because he didn't use those actual words — but there was no mistaking the implication: any consideration of the separation of the major units would be the end of modern GE.

Inside the company and on the board, it was clear that everything was now on the table. The insurance charge was crippling, and Trian was

calling for a complete reconsideration of all parts of the business in order to recover something for shareholders.

Flannery was constantly putting out fires, and then finding more fires. In addition, there was tension between the board and Flannery, as well as between Jamie Miller and Flannery.

Overall the directors supported Flannery as he worked to get the damage under control. In the minds of his supporters, Flannery was making all the right moves, even if a little more slowly than hoped for. He kept saying he just needed time. And with the board changes, the directors knew that new help was arriving to assist Flannery. After an extensive search, it had added three members who could all bring tools to solve GE's ongoing problems in its management and in its boardroom. But in the turnover, there was little love lost between the newer board members and the older ones.

GE's board of directors was unquestionably weakened from having the CEO as the chairman of the board. Many companies still cling to this structure, but it openly violates basic corporate governance principles. The CEO shouldn't also be in charge of the committee that employs him. Some GE directors would challenge Immelt, but that usually was fruitless and could land them in the penalty box, or sometimes off the board entirely.

Directors had often been chosen to join the GE board without a comprehensive search, or confirmation that they understood how the board worked. In addition, GE board meetings were often a whirlwind production. With eighteen directors and another dozen regular attendees, the room was packed, and it was impossible to tackle all of the vast number of issues on the agenda. In fact, executives and advisers had occasionally looked up to see some of the directors drifting off to sleep.

While Immelt said that he encouraged debate, meetings often lacked rigorous questioning. One executive recalled being in a board meeting in which Keith Sherin was presenting the quarterly financial results to the group. The Power business had missed badly, but little specific detail was provided on what went wrong. This executive braced for the reaction from the directors, but it never came — none of them asked what went wrong.

When Flannery committed to renewing and shrinking the board of directors, it included half a dozen current or former CEOs, the former

head of mutual fund giant Vanguard Group, the dean of New York University's business school, as well as a former chair of the Securities and Exchange Commission. The seventeen independent directors got a mix of cash, stock, and other perks worth more than $300,000 a year. The terms had been even more generous when GE still made appliances; the company allowed directors to take home up to $30,000 worth of GE products in any three-year period. The company matched the directors' gifts to charity, and upon leaving the board, a director could send $1 million in GE money to a charity.

Some directors admitted to having been sold by Immelt's sweeping optimism, even if they knew he wasn't the best deal-maker. But they knew he had a hard job, was playing with a tough hand, and had survived multiple major crises. Plus, they liked him.

Immelt said that he did his best to keep directors informed, noting that he required them to make trips to GE divisions on their own, but he also knew that the complexity of the business limited their input. As they'd done under Welch, the board usually tended to approve his recommendations and follow his lead.

Some felt that Immelt manipulated the board, and it was whispered that members were chosen and educated to see the company through his visionary eyes. There was concern that the board didn't entirely understand how GE worked, and that Immelt was just fine with that. Like many CEOs who are also their company's chairman, he made sure that his board was aligned with him.

In 2016, for instance, Immelt pushed out Sandy Warner, a twenty-four-year GE director and the former CEO of JPMorgan, after the two clashed over Immelt's succession plans. Warner thought that the succession should be speeded up, and that Steve Bolze, head of GE Power, was the man for the job. Immelt, at odds with Bolze and dissatisfied with his results at Power, felt that he had to force Warner off the board in order to torpedo Bolze's chances.

Warner appealed to his fellow directors in a closed session. He argued that they should at least allow a debate on whether it was time to replace the CEO. But the board stuck by Immelt. Warner left the GE board embittered after nearly a quarter-century. GE told investors in a securities filing that Warner left because of new term limits; the company never disclosed the real reason.

Immelt often made it a point to go around the board table to ensure that everyone had a chance to comment on a strategic decision, but directors rarely challenged him. To Immelt, this was proof that he solicited input and encouraged debate. But what the Federal Reserve had seen, when it was supervising the company, was a board that needed to push back more on the chairman. The Fed, including Immelt's nemesis Caroline Frawley, urged them to do so.

Just like Welch and Immelt, Flannery wanted to make the board his own, but he also wanted to strengthen its membership while shrinking its size. Flannery relied on the board for guidance; he didn't want people just rubber-stamping his decisions. He truly sought genuine debate, which was one of the reasons he welcomed Ed Garden onto the board.

Flannery also recruited Larry Culp. The former CEO of a smaller conglomerate called Danaher Corporation for fourteen years, Culp had run a tight ship when it came to deal-making and capital allocation. Danaher shares surged during his tenure, and he had retired at fifty-two, after making more than $300 million. The company sometimes came up at GE board meetings as an example of a conglomerate with a more streamlined and functional model.

Culp, a tall Maryland native with a full head of thick gray hair pushed to the side, wears suits that sometimes seem a size too large, speaks with a hint of a southern accent, and isn't easily flustered. His reputation had extracted respect from investors in industrial companies, but wider recognition was yet to come, partly because of his aversion to media attention.

In naming Culp to the board, Flannery was fully aware of the possible future that appointment might bring about. Before Culp joined the board in April, an adviser had warned Flannery that Culp would be the man to replace him atop GE if things soured. He didn't care, Flannery said. He just needed the best people to help him right the listing ship.

"MANAGING IN A BROADER SENSE"

I N LATE MAY, it was Flannery's turn to take the microphone at the Electric Product Group conference in Florida.

It had been a long year since Immelt delivered his less than stellar performance. All eyes were on the new CEO, with many of the same questions in mind: Was he going to deliver a new strategy? A major asset sale? A real solution to the GE Capital problem?

Flannery's reputation was beginning to harden among those at the top of GE. It was sensed that he lacked composure, that he could get flustered in high-pressure situations. His critics thought he sometimes flew off the handle. This was a surprising change in his behavior for many who knew him as a calm and confident leader.

Like Immelt, Flannery was a strong speaker. After the annual management retreat in Boca Raton back in 2015, Flannery's presentation had become part of GE lore. That annual three-day, invitation-only gathering in January was the ultimate networking event for GE leaders. Speakers sometimes included world leaders, and presenting to that group was a big deal. In Welch's day, preparations would begin before Christmas, with chosen executives presenting privately to Welch for advice and approval.

But in 2015 Flannery had stunned the room by pulling off a feat that, to GE veterans, was akin to walking on water: he presented without using PowerPoint slides. He had simply communicated what he felt the Healthcare business needed, including what it was doing wrong. That

singular triumphant moment helped build Flannery's reputation for self-confidence and a willingness to break from the pack.

But now, three years later, it was clear that being CEO of GE wasn't the same as presenting at Boca Raton, especially at a moment when everyone was looking to him to save the company. Flannery unfortunately was now building a new and different kind of reputation: for seeming to drop GE's stock price anytime he opened his mouth.

His handlers feverishly prepared him for EPG so as to avoid another setback. They gave him a long sheet of possible questions with appropriate answers, did mock sessions, and asked him the same questions in a variety of different ways so that he could always know how to steer to the best answer.

But despite all the extensive prep work, Flannery still came before the annual gathering with a gloomy and, to him, unavoidable message: the Power business was facing years of pressure ahead, and major changes at the conglomerate would take some time to show results.

When pressed, he declined to commit to GE's dividend for 2019 and answered questions about it as he typically did during his time at the top — as a finance expert. The dividend reflects the ability of the existing portfolio to pay it, he would say, so it might change with the portfolio. If a company sells half its businesses, he'd point out, it can't pay the same dividend.

What investors heard in this response was: I may cut the dividend again.

Flannery didn't follow the unwritten executive rule that you stand by your commitment — until you don't. In some ways, he was being too transparent. He refused to paint himself into a corner by making a long-term promise in a fluid situation. He also defended his methodical approach to the strategic review.

"So being deliberate and then moving when things make sense — as opposed to moving just because somebody wants us to — is just my style," he told the crowd. "So I get that people want faster. I'm managing in a broader sense."

GE's stock slid 7 percent that day on his comments. Again, the stock had dropped when Flannery spoke.

Even if the upbeat optimism of Immelt was now being seen as part of GE's troubled past, the former CEO's ability to communicate had at least

often calmed a room. The company and its army of managers could do anything, he convinced his listeners. It spanned the globe, was admired by world leaders, and could ultimately work out any problem thrown at it.

Immelt often lamented the hand he was dealt, the external catastrophes he had to navigate, but there had been one solid truth in his repetitive invocation of the past — GE had weathered a lot of storms with a business model that many thought was obsolete. And even some of Jack Welch's confidants had privately admitted that Welch couldn't stand up to Immelt in giving a speech, working a room, and handling a crowd.

For John Flannery, Immelt was proving to be a hard act to follow.

A MIDSTREAM CHANGE

I N JUNE CAME news that struck at the heart of the GE mythology: GE was being dumped from the Dow Jones Industrial Average.

General Electric was an original member of the thirty-stock index and had been included continuously since 1907. Its removal from the Dow sent a clear message: GE was no longer one of the most important US companies. It was replaced by Walgreens Boots Alliance, Inc., a drugstore chain and distributor that was half as big as GE in terms of market capitalization.

GE zealots were dismayed to be told they no longer worked at one of America's most prestigious companies. Flannery, on the other hand, took the news in stride, even though he knew that the GE rank and file would see this development as a setback. He knew that the move was probably inevitable, but also that the Dow wasn't as important as it once had been.

Right now, Flannery had bigger things on his mind. In a week, he was going to unveil a major plan to break up the company spawned by Edison and Morgan, then made and remade by Immelt, Welch, and the others who had come before him. The moment weighed on the GE veteran.

Flannery's preparations for his talk were going to be intense. Investment bankers, crisis PR firms, and other advisers were brought in. Several consultants were hired to help him deliver the message, which was

going to be done on a conference call, not in a presentation. The surprise announcement was set for June 26.

On the call, Flannery delivered the stark plan that many had expected back in November. GE would spin off its Healthcare division, sell its stake in Baker Hughes, cut its debt, and streamline its vast corporate structure. Once again, GE Capital remained unresolved.

Almost as an afterthought, Flannery announced that Larry Culp, who had joined the board earlier in the year, had been named GE's lead director, replacing former Vanguard boss Jack Brennan. Culp was the kind of operational expert that investors, including Trian, wanted steering the board.

Culp grabbed the reins in the summer board meetings, drilling Flannery with questions about the Power business and scolding the new CEO in front of the directors for not knowing such nitty-gritty details as inventory levels. Given the sprawl of GE, few expected Flannery to have such details at the ready, but not Culp. He was a former CEO who, at a much smaller business, had been known for immersing himself deeply in its various companies. Rather than bringing Danaher executives to headquarters for reviews, he would travel to their offices and walk the factory floors.

Flannery felt that he was bringing scrutiny to major issues that had been previously glossed over, like how to best spend GE's money. But for some on the board, Culp's dressing-down revealed a bigger problem. Flannery lacked the experience to juggle the steady flow of crises while also running a company that he was still learning about.

By this time, there was already a faction of the new GE board that wanted to consider a change in CEO. These members were concerned that Flannery wasn't up to the job and felt that the company shouldn't wait for a disaster before making a change. Over the summer, frustration mounted as Flannery's pattern of indecision continued. Then his support vanished a few months later when the board learned about new problems with the Power business.

Flannery had set up a conference call in September to brief directors on the source of the problems. The blades in GE Power's newest line of heavy-duty gas turbines were failing. Exelon, a big utility, was forced to shut two power plants in Texas for repairs, and GE would need to fix

dozens of other turbines it had sold that used the same blade technology. That promised to dampen already weak sales and drive up maintenance costs in the struggling unit. GE had been counting on that turbine to battle rivals such as Siemens.

There was more. GE was on pace to miss its cash flow targets and would have to take a charge of more than $20 billion to write off the value of previous acquisitions, including Alstom.

The board had had enough. They decided to hold Flannery accountable for the latest raft of bad news. After sixteen years of Immelt, Flannery thought that he had more time to turn the ship around, but when he looked for support from the board, there was none there.

By the end of the weekend, Flannery was out, fired after fourteen months, the shortest stay at the top in GE's long history. The new boss was Larry Culp.

Flannery was reeling. He had told the board that the turnaround would take time and be painful. And the board had pledged their support, then given up on him. The board in turn worried that Flannery was in over his head and knew they couldn't risk stumbling on the execution.

Culp was up for the job. Happily retired, he was teaching at Harvard Business School, where his students raved about his classes. He showed up for his class on the day he was named CEO to explain to them why he wouldn't be their professor anymore.

His time on the board had shown him that there was an opportunity for him to help fix GE. And he was a good fit — he had fourteen years of experience as a CEO, but was only fifty-five years old.

Culp knew that it could be a long slog, but he'd spent a lot of time thinking about whether he would take the job or even join the board. Culp didn't take the decision lightly. He had retired, but craved the chance to execute a gargantuan corporate turnaround. Besides, who could turn down the opportunity to save General Electric?

Unlike Flannery, Culp was aware of what he was walking into — at the very least he could brace himself before impact. Within weeks, he slashed the GE dividend again, leaving investors to collect a token 1 cent per quarter.

Culp also brought more credibility than Flannery, because of his CEO experience, and a new face to help the board forget the previous disas-

trous year. Still, despite the frustration with Flannery's methodical ways, Culp didn't actually approach the situation all that differently. He also planned to largely dismantle the company, which was still beset by investigations, lawsuits, and waning confidence that it could pay its debts.

Culp's declaration that the turnaround of GE was a multiyear project echoed Flannery's long-held view. Flannery's decision had been the right one.

Some of the major issues were still there, all impervious to a change in management. Federal criminal and civil investigators were looking into GE Power's modifications of service contracts to wring out more short-term profits. They were probing how GE Capital disclosed its continuing liability for long-term care insurance, as well as write-offs of Alstom and other deals.

Shareholders accused the company of defrauding them, citing the Power contracts and insurance liability in their lawsuits. GE has denied the allegations. And GE, once the owner of credit almost as good as that of the US government, saw rating agencies again drop the grades on its once-golden bonds.

And like Flannery, when Culp spoke, the stock sank. In a fidgety TV interview in November, he said that the Power business had yet to hit bottom. He declined to set new financial targets. GE's stock soon fell below $7 for the first time since the financial crisis in 2008.

The collapse was so complete that there was little left to lose. JP-Morgan analyst Steve Tusa, who had led the pack in arguing that GE was harboring serious problems, removed his sell rating on the stock in December. He carefully explained that GE's businesses were broken, but the risks were now mostly known. The stock climbed back above $7. It would take only a few months before Tusa was back on the attack, asserting that investors were letting the stock price get ahead of the turnaround. He again urged clients to sell GE shares.

As far and as hard as the fall looked to those on the outside, it felt even more so to those who had been on the inside, the true-believers like Jack Welch and Jeff Immelt and John Flannery. The pain was also shared by many of those lower in the company ranks — the shop-floor workers with a multigenerational legacy at GE. They also still believed, in GE management and in the products they made. To them, GE would never die.

All of these believers were trying to reconcile how a company valued at nearly $600 billion almost two decades earlier was now worth a fraction of that.

After being forced out, Flannery kept his distance from GE. At fifty-six, he set out on a six-week road trip with his wife, a journey he had long dreamed of taking. In three decades of climbing to the top of General Electric, he'd never before had the time. He crisscrossed the same American landscape that, not too long before, he had been cruising above in business jets. He felt burned but unrepentant. He did not regret refusing to be rushed in his attempt to salvage the company he loved.

Like so many others, he felt deep despair about its decline, people who know him said. He could have worked to keep everything together by any means possible, including patching up Power until it could be turned around rather than facing the problems directly. Instead, when Flannery saw the many broken sections of GE, he immediately moved to expose and fix them.

Culp had more experience, and he also had no emotional attachment to GE. Even if Flannery was facing the problems that Immelt wouldn't, or couldn't, face, many on the outside had worried that GE's problems were too serious for a former GE insider to handle. Flannery had often declared that there were no sacred cows, but that was a disclaimer Culp didn't need to give.

Flannery revealed serious problems, and every decision he made was a hard one, affecting thousands of factory jobs and lonely retirees waiting for a dividend check. In exile, Flannery remains certain that there was no quick fix to GE's problems, no matter how much investors and the board wanted one.

So far, he hasn't been wrong.

Immelt spent his first year outside of GE dividing his time between Silicon Valley and Boston. In California, he has embarked on the life of a venture capital partner, making tech investments and serving on startup boards. Back east, he served as chairman of Athenahealth, Inc., a medical software company. His office, in a redeveloped old arsenal that buzzed with tech companies, was sunny and modern, the antithesis of his old Fairfield corporate bunker.

GE still weighs on him. His anguished departure from the company to which he'd devoted his life remains fresh. He feels misunderstood and unfairly portrayed. Immelt still talks — too much, some of his allies say — about the missteps of others at GE, and why his way was right.

He talks of GE as if he still works there. He says that he wasn't imperial when he served as CEO, but refers to "my board" and "we" and "our" when talking about the company. He likes to remind people that he hasn't sold any of his GE shares. Don't bet against a comeback, Immelt says.

The former boss sees his GE tenure as an epic struggle — a Sisyphean effort to unwind the embrace of Capital that was thwarted over and over by shortsighted investors and pundits and by unanticipated failures in the Power business that rattled GE to its core.

"The notion of plugging financial services and industrial companies together, maybe it was a good idea at a point in time, but it is a uniquely bad idea now," Immelt said, more than a year after he stepped down as chairman.

Immelt admits a share of blame for GE's condition, but quickly turns to talking points about the righteousness of his strategy and his understanding of how to navigate organizations through complex and transformative times.

Just months after Immelt stepped down, and as GE shares continued to slide downhill, the *Harvard Business Review* published an essay by Immelt titled "How I Remade GE." The essay was a victory speech, one consistent with Immelt's approach over the years as he told readers how they should feel about his years at GE, using suggestive section titles like "Be All In" and "Be Willing to Pivot." "It will take years for GE to fully reap the benefits of the transformations," Immelt wrote. "I'm confident that I'm handing over a company that will flourish in the 21st century." At more than six thousand words, the essay was shocking to almost anyone who had a hint of GE's true condition.

Immelt seemed to be a man untroubled by the past while he adjusted to life in the wild, where he had to figure out, among other things, how to fly commercial again. Immelt told associates he paid no attention to the media that bedeviled him at GE. He also told them that he was at work on a book, his own version of the Jeff and GE story.

By later 2019, many current and former GE executives still awaited the outcome of federal inquiries into their departments' accounting and business methods. The uncomfortable cloud — and associated legal bills — still hung heavy over those who were sure they had violated no rules in running GE's books, which remained under scrutiny from the SEC and Justice Department. The government had collected a massive trove of emails from executives to scrutinize, causing considerable consternation inside and outside the company. The investigation was done only when the government decided it was done. Until then, they paid their lawyers and tried to focus on new ventures and hobbies.

Former GE Power CEO Steve Bolze landed at the private equity behemoth Blackstone, where he seemed essentially to disappear. Jeff Bornstein went on to make venture investments, having hung out a shingle in Ridgefield, Connecticut, back home in the state he had once helped GE storm out of in a huff over fiscal policy and politics.

Ed Garden remains on the GE board and occasionally makes public appearances to declare his faith in Culp's turnaround skills and his optimism for GE. Nelson Peltz told CNBC in late 2019 that Trian had made a "big mistake" by not selling all of its GE shares years earlier.

Almost to a person, these former GE executives now popped up on LinkedIn, the Facebook of the middle-aged corporate set, adding likes to their former colleagues' updates on new jobs or late-career affirmations of change. They share news of new ventures. Joe Mastrangelo, from Power, ended up in Italy. Bornstein showed a marked knowledge of the grow stages of a friend's cannabis business and doled out compliments to ex-GE finance pros as they moved on to new positions in other companies. Beth Comstock promoted her book *Imagine It Forward* and appeared in digestible and forgettable videos about notions like disruption and creativity and the power of "change."

John Flannery eventually shared that he had been hired by a private equity firm in Boston, having returned to the skill set he knew best — deal-making and advising, capital allocation, bluntly assessing what worked and what did not.

Some GE lifers saw Flannery as a tragic figure, taking the fall as he tried to clean up years' worth of others' mistakes. Flannery has told

friends that he remains certain that there was no quick fix for the serious problems he unearthed.

Slowed down now from the relentless pace of his glory years, Jack Welch continues to privately rage about the man he chose to succeed him. The aging CEO still likes to say he gives himself an A grade for the way he ran GE, and an F for his choice of successor. A conversation with Welch inevitably circles around to the latest news of the company that he managed into the stratosphere. Now he is livid at the collapse of the pivot, at Immelt's buybacks and spending, at the simple failure to run Power right.

Welch scolds the board for hanging on to Immelt too long. They could have replaced him long ago, Welch has argued, and had good reason to. "I'm terribly disappointed. I expected so much more," Welch noted. "I made the best choice I thought I could make, and it didn't turn out right."

Jack Welch doesn't talk about the old GE coming back to life, though he hasn't given up faith in the conglomerate model. He hopes that Larry Culp can "build a new GE" instead.

Culp faces the challenge of rescuing one of America's great companies — with decades of culture and pride behind it — as an outsider. He has made no secret that he is shifting the company's culture toward bottom-up management rather than imperial decrees from the top that don't resonate with those in the trenches. He is pushing the huge corporate umbrella organization to make managers more accountable for the business they run and attempting to get GE away from its decades-long end-of-quarter push to make numbers. Culp wants to create a "market for truth" in the company, an initiative that takes direct aim at the "success theater" under Immelt that Flannery had tried to fight.

Under Culp, GE has sold its transportation business, exited the oil and gas fiasco, and unloaded Healthcare's biopharma operation for $21 billion, in a move he says will help lighten GE's massive debt load. At this writing, the business most associated with GE's old avatar Thomas Edison, the lighting division, is still for sale.

Culp has pointed to the GE culture of Welch's days, when operations were focused and lean manufacturing honed the company's efficiency.

He is a student of the purest form of those processes, which came out of Toyota. The outsider CEO has tried to rally workers by declaring that he is one of them. But even the success of Culp's efforts would undermine a central tenet of GE's oldest and most precious belief: that it knew how to manage any business and could teach any of its own to do so.

In the end, when General Electric most desperately needed a manager to save what was left of the company, it had to go looking somewhere else.

EPILOGUE: "JEFF IS A FRIEND"

THE OBSERVATION BEGAN to seem like a refrain, volunteered in conversations over the course of months, then years, as those who had been close to the leaders of General Electric — current and former GE employees, bankers and lawyers, directors and analysts, shareholders and retirees — pondered and parsed the aftermath of its catastrophic attempt to change. Pressed to diagnose the causes of the company's failure and the motives behind the collapse of its plans, even the bitterest critics of GE's strategies tended to begin the crucial assessment of blame with the same instinctual caveat, the one thing on which many of them would ever agree: "Look," they would tell us over and over, "*Jeff is a friend.*"

The statement was incantatory and unconvincing, like politicians paying lip service to cooperation with opponents before savaging their positions on the Senate floor. But it was also true, in a sense. Jeff *was* a friend. The warm smile and chuckle, the piercing attention to life histories, the body language — these may have been simply the salesman's tools that he wielded with everyone he encountered in business, but it was also true that even some who fiercely disagreed with him believed that his warmth was genuine. Immelt was capable of small kindnesses, of convincing encouragement. He remembered his people and looked after them in unexpected ways. Supportive attention to the plight of children, collegial ribbing about a football team, an invitation to come up to New Canaan for a weekend at the big house, a few rounds of golf.

But in truth, he could also be cold. Joshua Raymond was packing up

his office in GE's Washington headquarters in the late summer of 2016. With GE getting out of lending and Congress in Republican hands, Raymond, a former Democratic congressional staffer and an expert in financial and regulatory matters, had become expendable. There were no hard feelings. Part of the business of Washington was understanding the cycles and knowing when it was time to move on.

Before Raymond's last day, Immelt made an appearance in the company's large Washington office. The CEO strode down the hallway, passing within earshot of other company executives, including Nancy Dorn, the company's fearsome head of government affairs, when Immelt spied Raymond.

"Why are you still here?" the CEO sternly asked. The others recoiled as Raymond, his brow furrowed, mustered a response: "I'm still here for two more weeks."

"Oh," Jeff Immelt said brusquely, then walked away.

In the weeks and months after Immelt left GE in 2017, a parade of negative stories and embarrassing disclosures revealed major problems that sent the company's stock into a long decline. Conversations about what happened inevitably shifted to blame, and Immelt was the obvious target. He had spent sixteen years at the top and, regardless of what Welch had left for him, he'd had plenty of time to fix it.

But there was plenty of blame to go around. Perhaps most of it should be placed on the board of directors, the independent group that oversees the CEO. Board members claimed to have been unaware of problems and to have gotten bad guidance from external advisers, and they said they didn't understand how the company went from good to bad seemingly overnight. Some directors had no experience in GE's business lines, others had trouble staying awake during meetings, and many stumbled away from GE's collapse wondering, *How could we have known?* It had been their job to know, however, and their job to ask the hard questions that weren't fully answered, or were never asked at all. It was their job to oversee management, and it was their job to protect investors from fatal hubris.

Still, the path ultimately leads back to Immelt. As chairman, he was also responsible for steering the board. There is no doubt that GE's size and complexity, which grew exponentially under Immelt, made it dif-

ficult or even impossible to manage. The CEO of a company is responsible for its daily functions and for managing its operations, however vast. The chairman guides the board, which is responsible for overseeing management and the CEO. When the board chair and CEO are the same person, the top executive is essentially his own boss. It can only get worse with time if a chairman remakes the board to his own liking. Simply put, it is terrible governance to give so much power to a single person and so little voice to shareholders. That is one reason this governance structure has been slowly fading from corporate America since the Enron era.

There is also plenty of blame to put on GE's top-down culture, which Welch and any number of midlevel managers used to their advantage as readily as Immelt did. They instilled a by-any-means-possible mentality and a lax attitude toward costs in an organization so gargantuan that mistakes were easily hidden. Unleashed on a grand scale year after year, these practices are inevitably crippling. In effect, GE's size advantage was also one of its biggest weaknesses.

An assessment of Immelt using the stark metrics of share price or investors' returns easily paints a dismal picture. But such a simplistic judgment ignores his leadership, his successes, and his efforts to make GE more diverse and innovative and known for something other than a rising share price.

Immelt had a bold vision, but ironically, the corporate leaders he idolized were decidedly not cut from the same cloth as GE people. He admired company founders like Amazon's Jeff Bezos and FedEx's Fred Smith, both of whom had an idea, established a company, and then managed its growth to successfully outrun competitors. Perhaps it is understandable that this vision would be appealing to a leader who came into the job after eighteen years working for one of the most famous CEOs in modern times and whose success would be attributed to his predecessor but whose failure would be seen as all his own doing.

Many feel that Immelt succeeded in his effort to transform GE. His supporters insist that history will show the brilliance of Immelt's vision for the company. Taking a position that is tough to defend, some argue that GE was actually on the correct path when he resigned and that dismantling his strategy caused the collapse, not the strategy itself.

Immelt's departure was cause for celebration for some inside the

company. Others, his closest aides and loyalists, shed tears. The day after Immelt's departure broke, word reached the company communications office that the *Wall Street Journal* was working on an "ahed" — the lighthearted front-page story that had been a *Journal* feature for decades and served as a venue for a gentle ribbing — about Immelt's legendary insistence on utterly frigid air conditioning in his public appearances. If nothing else, the *Journal* suggested, his resignation might allow company employees and observers to leave their sweaters at home.

Immelt's predilection for cold was actually humanizing — even endearing to some who disdained him, including some of the analysts who loathed having to wear their Fairfield County barn coats to the annual investors' meeting at 30 Rock. Immelt's brother Stephen had also cheerfully cooperated with the story, noting that his brother was going hog wild for chilly air conditioning, as he had done when it had been a sought-after luxury in their childhood home in southern Ohio.

Lighthearted the *Journal* piece might have been, but one of Immelt's closest allies went ballistic.

"This is a man that has literally sat with kings and world leaders representing this company and our country," said Deirdre Latour, then GE's chief communications officer, in an email fired off to reporters and editors at the paper. "This is a man whose driver and security guard were crying yesterday" when they heard about his resignation, because, she said, "he treats people with the greatest respect."

Immelt had "inherited a total mess from Jack Welch," Latour declared. Echoing the language of a newly inaugurated president, she claimed that Immelt had "made this company great again." "It is offensive to me that the WSJ would do this as a day 2 story about one of the most consequential American business leaders in history," Latour wrote. "Disgraceful."

But Latour's email blast failed, and the ahed ran. As overwrought as her defense of Immelt seemed, she had a point. Jeff Immelt had sat with kings and was indeed a "consequential figure in American business." Looking at GE through the lens of the stock price alone was simplistic. The portfolio had been as transformed under Immelt, by some measures, as it had been under Welch. Immelt *was* trying to make GE great again.

Latour didn't want a cute story about air conditioning. She wanted

the *Journal* to finally write the story that she and GE's press operation had demanded for years: a sweeping examination of Jeff Immelt's legacy — the ways in which he had changed the company and thereby changed the trajectory of business in America and around the world. "I get that the day 1 [story] has to be about the numbers and performance," Latour wrote. "It is business journalism. This as a day 2 is wrong." Furious, the spokeswoman banged out the short, deceptively simple question that she felt the *Journal* needed to answer if it was going to give the retiring CEO the send-off he deserved: "What did Immelt actually mean?"

More than a year later, a former GE official learned over a breakfast in New York that Immelt, out of power and wincing at the drubbing he had taken in the markets and the press, was heaping blame on, of all people, Deirdre Latour. She was an unwaveringly loyal defender of his image, and though she may have sometimes gone a bit beyond reason in her dealings with the press, Latour seemed an odd target for the CEO of GE to be blaming for his downfall. The official — mentioning, naturally, that Jeff was a friend — shrugged at the notion of the boss casting an aide so casually aside: "Sounds like Jeff."

Still, Immelt's aura was embracing where his legendary predecessor's had been piercing. Competitiveness and passion came off each man in waves, though tuned at different frequencies.

"Jeff is a friend," his contemporaries would say. Then they would explain how his warmth and optimism had hidden shortcomings that undermined the processes and deliberations that were supposed to be GE's hallmark — how the blind spots of the biggest believer in the prowess of the company had ruined its impregnable image and helped throw its entire existence into question.

GE people have found themselves forced to explain the past, to assign measures of blame while attempting to make sense of it all. Some are simply defensive. Others are seething with anger. Some won't talk.

And sometimes the effort to defend the company's performance has veered into the absurd. Months after Immelt's resignation and Flannery's firing, one former director defended one Immelt-era gambit: the decision to plunge into the oil and gas industry, a move that by this

point looked to most observers like a classic example of GE fad-surf-ing. They had overpaid for businesses they didn't understand and then been crushed by the market. Though not defending Immelt, the direc-tor bristled at the notion that GE's board hadn't been policing the CEO's deal-making.

This criticism had been the muttered verdict on big company boards for a generation, various reforms notwithstanding. But as the ranks of conglomerates dwindled, especially any that could rival GE's breadth and complexity, the need to believe in an inquisitive and judicious board grew more and more compelling. GE's rhetoric about management and its own self-evident belief in its talent-grooming notwithstanding, a gi-ant public company exposed to economic risks as divergent as the price of a barrel of oil, the fashions in home appliances, and the air travel growth rates of the mega-economies of Asia required, at a minimum, a serious and fully engaged board of directors.

So the implicit argument went anyway. If Immelt felt burned by the sudden consensus that he had been a leader betrayed by optimism and oblivious to unpleasant truths — that he was a visionary fool — so too did the board now recognize an unflattering portrait of itself. The di-rectors had amassed impressive titles in their own careers and in many cases undeniable achievements. They had résumés a yard long, most of them had personal fortunes, and they were presumed in all company to have unusually astute minds for business — not least because each one was a highly compensated director of General Electric. And yet, on their fiduciary watch, with whatever caveats about individual misjudgment and macroeconomic trends, they had done nothing to stop one of the world's most solid industrial companies from plunging itself off a com-mercial cliff.

Multiple directors assert that they scrutinized GE executives closely. They emphasize that the board had members who knew the company's diverse industries well — even oil and gas, which looked in hindsight like a business GE had gone into cavalierly, with no decent exit strategy. But then, with Jim Mulva, the former CEO of oil giant ConocoPhillips and a man with a broad understanding of the industry and its risks, on its board, why was GE caught so flat-footed by declining oil prices that slowly undercut Immelt's rationale for bulking up in the oil equipment

business? In response, one director sputtered in indignation. Mulva had a sharp mind for the industry, he replied, but added, "He doesn't hear so well."

When it came, GE's once unimaginable decline happened quickly. In the year after Immelt left, more than $140 billion in stock market value evaporated from the company. More than twice the amount that vanished when Enron collapsed in 2001, that loss dwarfed the market losses on Lehman Brothers.

The rapid unraveling didn't just hit big Wall Street firms. Long one of the most widely held US stocks, shares of GE sit in the portfolios of hundreds of thousands of ordinary people who squirrel money away into the stock market for retirement, as well as the portfolios of legions of former GE workers, who were encouraged to buy company stock through subsidized purchase plans.

With equity incentives and stock option awards retaining little value, GE executives also took major losses. That damage was severe, though many of them benefited from ample compensation and lavish perks earned during their careers. They used GE jets to visit far-flung customers and company facilities, but some also used them to fly with their families on vacation. Top executives got cars fully paid for by GE, a program started by Welch for about 125 executives; Immelt expanded the perk to cover another 700. (It fell to John Flannery to kill the program.)

And then there was the pay. Welch was paid handsomely, partly because of his success and the skyrocketing stock, but he also locked in perks that would continue until after retirement, including a Manhattan apartment paid for by GE, access to corporate planes, sports and opera tickets, fresh flowers, and a personal chef. Welch himself would later surrender many of those perks when their disclosure in his divorce outraged the public and embarrassed the company. Nevertheless, some estimate that his wealth from working at GE climbed above $1 billion; the *Wall Street Journal* put it between $450 million and $800 million in 2002. His compensation for his last year at the company was valued at $122.5 million, the paper reported.

Executive compensation is difficult to measure because disclosure requirements weren't in place until 2006. There is no doubt that Jeff Im-

melt also enjoyed plenty of perks, including heavy use of the corporate jets. As with many things around Immelt, there is contention around his enjoyment of perks versus him needing the required tools of an impossible job that never slowed down for sixteen years. Some claim he would eat only a chicken salad sandwich on the corporate jet; others claim the crew had to stock both cooked lobster and steak so Immelt had a choice while in flight. Some of these questions may never be settled.

Despite the dismal performance of the stock over his tenure, Immelt was paid handsomely, getting an estimated $168 million between 2006 and 2017 alone, according to the data firm Equilar. Jeff Bornstein's four years as CFO brought an estimated $37 million, according to Equilar, while Keith Sherin collected compensation valued at $108 million from 2006 to 2016. That didn't include Sherin's time in the Welch years or his service as CFO since 1998.

In a modern corporate world where executives are regularly paid huge sums to lead companies, the payouts at GE are still jarring. What was left for retail shareholders — the "widows and orphans" — and the GE retirees? The value of their stakes in the company has in some cases been literally decimated. And the dividend checks on which so many expected to rely in their later years will now be counted not in dollars but in pennies.

But time has passed GE by. It will be a long road back to the company that had hordes of recent graduates applying for coveted internships and job applications from those hoping to follow in the footsteps of previous generations of their family who had thrived on the shop floor.

One former GE executive recently said that his daughter hadn't even considered working at GE at the beginning of her professional career. "Are they even in the conversation anymore?" he asked.

In the end, Deirdre Latour had been asking the right question: what did Jeff Immelt — his tenure, his tactics, and his track record — actually mean? For that matter, what does GE itself mean now — to its stakeholders and to the society it helped to define? After all, any number of companies could sell machines to make power from natural gas or thrust American fighter planes up over open water.

The GE of the late twentieth century and the fitful first two decades of the twenty-first has stood for more than what it made. It has represented a capitalistic meritocracy, a locus not just of success but of a certain ver-

sion of virtue — the virtue of targets made, goals surpassed, earnings earned, markets won. And it has stood for a vague but well-marketed notion that, in the unapologetic pursuit of a company's fortunes, and one's own, there is a certain uprightness — and a lesson for others.

But GE has stood for a well-bred hubris as well. Under Immelt, the company believed that the will to hit a target could supersede the math, even when hundreds of thousands of livelihoods — those of investors, customers, and suppliers, to say nothing of workers, retirees, and their families — hung in the balance. GE became synonymous with the idea that all that stood between a company and an arbitrary $2 earnings target was drive — simply wanting it bad enough.

Here the company's aggressive accounting, macho culture, and reliance on competitive metaphor converged. Immelt's greatest asset as a manager, even in his staunchest allies' telling, came down to a football coach's insistence on the prevailing power of gumption and guts. Even numbers were supposed to give way, like a tackle sled on a football practice field, if you cared enough, lowered your shoulder, and wanted it more than the next guy in line.

Where, Jeff Immelt used to ask from behind the dais in the Pit, is the guy I used to know? Your people don't want it bad enough.

This was inspirational leadership — right up until it wasn't.

Reality isn't a tackle sled. Story isn't a strategy. Sometimes imagination alone doesn't work.

ACKNOWLEDGMENTS

We couldn't have written this book without the *Wall Street Journal*. Ted started covering GE for the paper in 2014, when the corporate bureau took a chance on a metro reporter who'd never even listened to an earnings call and let him cover one of the country's most prominent industrial companies. When Tom succeeded Ted in the spring of 2017, he met Jeff Immelt for the first time just ten weeks before the latter resigned as CEO. This book tells the story of the months and years that led up to that moment.

The *Journal* newsroom is a place of enthusiastic collaboration, unstinting dedication to balance, and cherished integrity. In the corporate bureau in New York and in Washington, we were given the freedom to chase the most important stories and break old molds in how we delivered them to readers. We benefited at every turn from the trust and deep commitment to the story of General Electric and its people on the part of our editors. Chief among them were Jamie Heller, Paul Beckett, Karen Pensiero, and Matt Murray.

The people of General Electric, past and present, were our greatest source of information, our best check on assumptions, and our reminder that the dollars and cents of this corporate story had real human consequences. As a rule, GE people love their company, and almost to a person they believe that the company's essence resides in its rank-and-file workforce, whatever the successes or failings of upper management. They ache for the company's revival. We are especially indebted to those

who spoke to us in spite of the pain they felt at reliving a disappointing period in the history of the company they love.

Countless others have contributed to our coverage over the years and helped make this book possible, including an army of people behind the scenes. Our colleagues were a source of constant support and encouragement and their contributions are too numerous to list in full. Among the *Wall Street Journal* journalists who played essential roles in telling the GE story were Marcelo Prince, Andrew Dowell, Kate Linebaugh, Joann S. Lublin, David Benoit, The Danas (Cimilluca and Mattioli) of the fearsome deals team, and Matthew Rose.

We are grateful to others at the paper who provided critical support: Paul Vigna, for his unending encouragement and coaching; Alex Martin, who showed us how to tell the story, put in painstaking hours and contributed his vertiginous vocabulary (and some spackle); Eric Lupfer, our agent, for his persistence, wise guidance on structure, and generous encouragement when we struggled to find our way; and Rick Wolff, our editor, for his belief in the story, steady hand and eye, and positive reactions to nervously submitted drafts. We are deeply indebted as well to Olivia Bartz in HMH Editorial for her efforts on the book, as well as to Loren Isenberg for her careful and meticulous legal review. Thanks also to Laura Brady, Lisa Glover, Cindy Buck, and Katie Kimmerer for their hard work on the production of the manuscript, and also a big shout-out to Emma Gordon for directing the publicity of *Lights Out*.

No one helped us find this story more than our direct editors on our coverage of General Electric. Andrew Dowell, Marcelo Prince, and Kate Linebaugh were the gatekeepers who decided when a raw observation was a story, and our defenders when the story brought on the substantial wrath of corporate PR.

Drew is a dynamo: fair and scrupulous, but unflinching, unafraid, and unflagging. He gave every reporter with a half-baked idea the command he or she wanted and dreaded — "just start writing" — then stayed with us to the last to ensure that the resulting story was smart, tough, and true.

Marcelo's patience, wisdom, and unyielding belief in our abilities was fuel to our efforts, and his quick eye on deadlines saved us many times. Unflappable in the face of pressure, he always gave us free rein to do the

job we needed to, secure in the knowledge that he would help us keep from tripping ourselves.

As a former occupant of the beat we shared, Kate Linebaugh knew GE as well as either of us. She couldn't have been more generous, whether with guidance about the company or her willingness to dive into any story to lend a hand. And she was blunt when she thought we had it wrong. Her friendship, leadership, and tough love were, and are, invaluable.

Tom: I want to thank everyone who helped me walk a rather unorthodox path through this world, beginning with the loving foundation and continuing support provided by my parents and brother Jay. I'm grateful to my in-laws for their assistance over the years. My wife Virginia provided unending encouragement and love through this process; she is my secret ingredient in life. Our children Maggie, Henry, and Frannie are the light of our lives. They keep me sane and smiling.

Ted: I'm lucky to be the child and the sibling of readers, and grateful to my family for their support in all things, above and beyond how they helped us write this book. I am even more fortunate to have in-laws, and a grandfather-in-law, who have been eager to help shoulder domestic responsibilities as I disappeared into the library for hours. Mostly, I am grateful for my wife Annie and my son Caleb. I couldn't have done this without their love, and I would not be here without them.

NOTES

A Note on Sourcing

GE is a conservative and hierarchical institution, with a culture that is innately resistant to scrutiny by outsiders, especially journalists. This book is the result of six years of combined reporting on General Electric, its business, its promises, and its people. In the course of our reporting on GE for the *Wall Street Journal* and this book, we have conducted hundreds of interviews with scores of people involved with the company — executives, members of the board of directors, bankers, lawyers, consultants, investors, and employees. We have spoken to some of the most prominent names in this narrative; we have also received valuable insights from people who are, as one GE executive derisively put it, "down on F deck" — those workers deep in the rank and file of the corporation whose names are not known to the general public but whose perspectives on the company have been instrumental in helping us learn about what GE did well, what it got wrong, and how those in the middle of the organization first spotted the signs of trouble.

Quotations that aren't otherwise credited are from interviews. Some were on the record, and some who spoke with us would do so only on background, providing information and insight on the grounds that we not reveal their identities or, in some cases, quote them directly. Where we have relied on background interviews in this book, we have operated as we would at the *Journal*. Details provided anonymously have been

cross-checked against public records and interviews with other reliable sources and across multiple conversations. In certain cases, we have used pseudonyms to protect the identities of subjects and sources. We have taken into account the unreliability of memory, especially as some events have receded into the past. And we have remained attentive to the motives of those we interviewed — both on background and on the record — in their framing of the events that befell General Electric.

I. OFF A CLIFF

page

4 *Flannery had taken:* Thomas Gryta, Joann S. Lublin, and David Benoit, "How Jeffrey Immelt's 'Success Theater' Masked the Rot at GE," *Wall Street Journal*, February 21, 2018.

2. THE MEATBALL

9 *One estimate pegged:* Brand Finance, "Engineering & Construction 50 2019: The Annual Report on the Most Valuable and Strongest Engineering & Construction Brands," July 2019, https://brandirectory.com/reports/engineering -construction-50-2019, accessed October 12, 2019.

11 *Edison sold his last GE shares:* Paul Israel, *Edison: A Life of Invention* (New York: John Wiley & Sons, 2000), p. 354.
 Besides frequent appearances: Julia Kirk Blackwelder, *Electric City: General Electric in Schenectady* (College Station: Texas A&M University Press, 2014), p. 18.

12 *career was on the wane:* Thomas W. Evans, *The Education of Ronald Reagan: The General Electric Years and the Untold Story of His Conversion to Conservatism* (New York: Columbia University Press, 2006), pp. 22–23.
 GE also led its peers: Ibid., pp. 45, 46.

3. NEUTRON JACK

14 *The most influential CEO:* Jack Welch, with John A. Byrne, *Jack: Straight from the Gut* (New York: Warner Business Books, 2001), p. 7.
 It was from Grace Welch: Ibid.
 Grace Welch taught: Ibid.

15 *"radiating noise and energy":* Bill Lane, *Jacked Up: The Inside Story of How Jack Welch Talked GE into Becoming the World's Greatest Company* (New York: McGraw-Hill, 2008), p. 23.
 Jones was the quintessential CEO: Matt Murray, "Why Jack Welch's Leadership Matters to Businesses World-Wide," *Wall Street Journal*, September 5, 2001.

16 *Welch shunned:* Ibid.

17 *During one visit:* Welch and Byrne, *Jack,* p. 96.

 Under Welch: Murray, "Why Jack Welch's Leadership Matters."

 By 1985: Thomas J. Lueck, "Why Jack Welch Is Changing GE," *New York Times,* May 5, 1985.

22 *wanted to take a different approach:* Welch and Byrne, *Jack,* p. 408.

23 *The directors spent time:* Geoffrey Colvin, "Changing of the Guard," *Fortune,* January 8, 2001.

4. BIG JEFF

24 *But then the compressors:* David Magee, *Jeff Immelt and the New GE Way: Innovation, Transformation, and Winning in the 21st Century* (New York: McGraw-Hill, 2009), p. 18.

25 *Immelt was often riding along:* Ibid., pp. 20–23.

26 *His older brother:* Adam Bryant, "Stephen J. Immelt of Hogan Lovells: Raised to Believe 'You Can Fix This,'" *New York Times,* March 14, 2005.

 Immelt himself: Claudia H. Deutsch, "GE's New Corporate Face; Jeffrey Immelt Rides a Can-Do Confidence to the Top," *New York Times,* December 1, 2000.

28 *What exactly happened:* John A. Byrne, "The Fast Company Interview: Jeff Immelt," *Fast Company,* July 1, 2005.

 "Jeff, I'm your biggest fan": Welch and Byrne, *Jack,* p. 421.

 "If the results": Ibid.

 Welch's wasn't an idle threat: Thomas Black, "Passed over for GE CEO, Dave Cote Thrives at Honeywell," *Bloomberg,* May 2, 2013.

30 *The flip side:* Byrne, "The Fast Company Interview."

5. THE RACE

33 *Welch was not impressed:* General Electric 1999 annual report.

35 *Like Immelt:* Bill Rigby, "Boeing CEO McNerney Gambles on Strike," *Reuters,* September 14, 2008.

36 *A* New York Times *profile:* Claudia H. Deutsch, "GE's New Corporate Face; Jeffrey Immelt Rides a Can-Do Confidence to the Top," *New York Times,* December 1, 2000.

37 *even used a contract plane:* Welch and Byrne, *Jack,* p. 424.

6. MAKE SOME NOISE

40 *Jeff Immelt's first day:* Joe Manning, "New Head of GE Gets Pep-Rally Style Welcome from Division He Once Led," *Milwaukee Journal Sentinel,* September 11, 2001.

41 *The previous Friday:* Lisa Marsh, "Jack Welch, Party Animal," *New York Post,* September 2, 2001.

The New York Stock Exchange: Ibid.

42 *That moment:* Jeff Zaleski, "*PW* Talks with Jack Welch," *Publishers Weekly,* February 25, 2005.

43 *The new CEO:* Diane Brady, "Jeff Immelt on His First Days Running General Electric," *Business Week,* September 1, 2011.
fresh-cut flowers: Matt Murray, Rachel Emma Silverman, and Carol Hymowitz, "GE's Jack Welch Meets Match in Divorce Court," *Wall Street Journal,* November 27, 2002.

45 *"We listened to everyone":* Beth Comstock, *Imagine It Forward: Courage, Creativity, and the Power of Change* (New York: Crown/Currency, 2018).

46 *"My second day":* "Perspectives," *Newsweek,* September 30, 2001.

7. THE EDISON CONDUIT

49 *"We think consistency":* Justin Fox, "Learn to Play the Earnings Game," *Fortune,* March 31, 1997.
"We're a very complex": Ibid.

53 *"a pretty consistent record":* Ibid.
Fortune's *Carol Loomis:* Carol J. Loomis, "My 51 Years (and Counting) at *Fortune," Fortune,* September 19, 2005.

54 *(SEC), Arthur Levitt:* Arthur Levitt, "Numbers Game," speech delivered at New York University Center for Law and Business, September 28, 1998.

8. SENIORITIS

56 *company had enjoyed broad trust:* Robert A. Bennet, "IR Goes Better with Coke," *Investor Relations,* May 1, 1996.

58 *"If the annual report":* Rachel Emma Silverman, "GE to Change Disclosure Practices to Include Details on GE Capital," *Wall Street Journal,* February 20, 2002.
Gross's research notes: William H. Gross, "Buffetting Corporate America," *Watching the World Tick Away,* March 2002.

59 *"GE has been shrouded":* Ibid.

9. LAST CALL

64 *The biggest boost:* Steve Maich, "GE Earnings Quality in Question: Too Many Lucky Breaks," *National Post,* October 1, 2002.

65 *Even large investors:* Alicia Seow-Suyama, "Calpers to Trim Bonds, Challenge GE over Exec. Pay," *Reuters,* October 15, 2002.

10. BUYING AND SELLING

67 *the turbo-charged deal-making:* "The Hard Way," *Economist,* October 16, 2003.

69 *Immelt had to plow:* Jonathan R. Laing, "Jack's Magic," *Barron's*, December 26, 2005.

 Welch had snapped up: Jack Welch, "Barron's Mailbag," *Barron's*, January 2, 2006.

70 *"Each quarter":* Ibid.

11. IMAGINATION AT WORK

72 *Beth Comstock had spent her career:* Comstock, *Imagine It Forward*, p. 113.

73 *claims to have been the one:* Eric Bloom, a guitarist for the band, also claims to have played the cowbell on "Don't Fear the Reaper," at Lucas's suggestion; see Tony Sauro, "Blue Oyster Cult's Innovative Use of a Cowbell Will Never Be Forgotten," *The Record* (Stockton, CA), September 17, 2009.

 "Don't Fear the Reaper": Joe Arena, "Blue Oyster Cult Cowbell Ringer Honored," WIVB.com, June 30, 2011, https://web.archive.org/web/20110702084549/; http://www.wivb.com/dpp/entertainment/music/Blue-Oyster-Cult-cowbell-ringer-honored (accessed October 13, 2019).

74 *"essentially means using dreams":* General Electric 1993 annual report, p. 5.

 "It's a reason for being": Wolff Olins Ltd., "Case Studies — GE," https://www.wolffolins.com/case-studies/ge/ (accessed October 13, 2019). In a case study of past client work on its website, which we accessed and archived in September 2019, Wolff Olins extensively described its work on the GE rebranding. The case study included a quote attributed to Immelt: "Imagination at work is not a tagline, it's a reason for being." As of December 2019, after we reached out for comment on the case study, the quote from Immelt had been deleted from the case study on Wolff Olins's homepage.

 company's 3,500 business offerings: Ibid.

 Never prone to underselling: General Electric, "GE.com/brand," www.ge.com/brand/ (accessed October 13, 2019).

75 *"My job is to make the company":* Rick Barrett and Thomas Content, "General Electric Boss Immelt Discusses Company's Recent Challenges," *Milwaukee Journal Sentinel*, April 21, 2002.

 Immelt's inspiration: Comstock, *Imagine It Forward*, pp. 119–20.

76 *"That turned out":* Thomas A. Stewart, "Growth as a Process," *Harvard Business Review*, June 2006.

78 *there were signs:* Charles Hutzler and Neil King Jr., "China's Premier Makes Appeal to Corporate US," *Wall Street Journal*, December 9, 2003.

79 *"In my view":* John Christoffersen, "GE Touts Products to Chinese Premier," *Associated Press*, December 8, 2003.

12. TIME TO GO BIG

81 *"No company has ever achieved":* Thomas A. Stewart, "Growth as a Process," *Harvard Business Review*, June 2006.

86 *They went, eventually:* "S&P Affirms Ratings on St. George and Superbank Post-sale," *Reuters,* August 3, 2006.

13. HIGHER RETURNS

88 *Some other pieces:* General Electric, "GE Focuses on Growth and Higher Returns with Sale of GE Advanced Materials Business to Apollo Management, LP," press release, September 14, 2006.
89 *In an early intracompany experiment:* Matt Murray, "Late to the Web, GE Now Views Internet as Key to New Growth," *Wall Street Journal,* June 22, 1999.
 Immelt said that the sale: General Electric, "GE Announces Sale of Plastics Business to SABIC for $11.6 Billion," press release, May 21, 2007.

14. APPLIED MATHEMATICS

92 *GE's practice of issuing: Securities and Exchange Commission v. General Electric Co.,* complaint, August 4, 2009. The SEC's investigation began in 2007 and the complaint was filed simultaneously with the 2009 settlement. The court filing includes extensive details of the investigation's findings. GE said it had already fixed the accounting problems and agreed to pay a $50 million fine. GE did not admit any wrongdoing. The company told investors it spent about $200 million in legal and accounting fees related to the investigation.
93 *The investigators found:* Ibid.
 In a much bigger mess: Ibid.
 GE's own accountants: Ibid.
 also a separate problem: Ibid.
94 *It isn't hard to see:* Ibid.
 A little imagination: Ibid.
 the $156 million difference: Ibid.
95 *"Our financial services businesses":* General Electric 2007 annual report.
97 *"financial services businesses should do well":* Ibid.

15. SIZE DOESN'T MATTER

98 *In one early passage:* Lane, *Jacked Up,* p. 36.
99 *His Boswell reports:* Ibid., 36.
 GE Capital saw: US Department of Justice, "General Electric Agrees to Pay $1.5 Billion Penalty for Alleged Misrepresentations Concerning Subprime Loans Included in Residential Mortgage-Backed Securities," press release, April 12, 2019.
 Like many companies: Ibid.
100 *Lending in the WMC division: Law Debenture Trust Company of New York v. WMC Mortgage,* complaint, October 26, 2012.
 One applicant's occupation: Ibid.

Every appearance: Ibid.

Some banks: US Department of Justice, "General Electric Agrees to Pay $1.5 Billion Penalty."

This behavior: Ibid.

101 "*I just think*": General Electric second-quarter conference call, July 13, 2007.

Welch was aware of: Loomis, "My 51 Years (and Counting) at *Fortune.*"

104 "*Even with land prices*": Kathleen Chu, "General Electric Japanese Property Investment May Reach 1 Trillion Yen," *Bloomberg,* August 25, 2007.

16. THE SCREWUP

107 *Comstock had struggled:* Meg James, "NBC Universal Executive to Move Back to Parent GE," *Los Angeles Times,* March 4, 2008.

108 *The market turmoil:* General Electric first-quarter conference call, April 11, 2008.

18. PAPER JAM

114 *Immelt had made a call:* Henry M. Paulson Jr., *On the Brink: Inside the Race to Stop the Collapse of the Global Financial System* (New York: Business Plus, 2010), pp. 227–228.

116 *That Monday night:* "Geithner's Calendar at the New York Fed," *New York Times,* n.d., https://www.nytimes.com/interactive/projects/documents/geithner-schedule-new-york-fed.

117 *Commercial paper was not:* Goldman Sachs, "Entrepreneurialism and Grit Inspire Marcus Goldman to Launch His Business," https://www.goldmansachs.com/our-firm/history/moments/1869-founding-of-gs.html.

In their research: Marcin Kacperczyk and Philipp Schnabl, "When Safe Proved Risky: Commercial Paper During the Financial Crisis of 2007–2009," *Journal of Economic Perspectives* (Winter 2010): 29–50.

The most notable impact: Ibid.

118 *On the morning:* Andrew Ross Sorkin, *Too Big to Fail: The Inside Story of How Wall Street and Washington Fought to Save the Financial System — and Themselves* (New York: Viking Penguin, 2009).

119 *The afternoon:* "Geithner's Calendar at the New York Fed," *New York Times.*

19. SUSPENDERS ON SUSPENDERS

122 *strong signal to the market:* James B. Stewart, "Eight Days: The Battle to Save the American Financial System," *The New Yorker,* September 14, 2009.

123 "*Even with all this volatility*": General Electric third-quarter conference call, October 10, 2008.

124 *Paulson called Immelt:* Paulson, *On the Brink,* p. 356.

"Maybe a lot": Ibid.

125 *In her memoir:* Sheila Bair, *Bull by the Horns: Fighting to Save Main Street from Wall Street and Wall Street from Itself* (New York: Free Press, 2012), p. 118. *"I decided to approve it":* Ibid.

126 *After repeatedly asserting:* Federal Deposit Insurance Corporation, *Crisis and Response: An FDIC History, 2008–2013* (Washington, DC: FDIC, 2017).

20. CIRCLE THE WAGONS

127 *"anyone could have run GE":* Steve Clemons, "The Worst of Times?," *Financial Times,* February 18, 2009.

128 *"We've got the cash flow":* Paul Glader, Eleanor Laise, and E. S. Browning, "GE Joins Parade of Deep Dividend Cuts," *Wall Street Journal,* February 28, 2009.

129 *columnist Joe Nocera:* Joe Nocera, "Behind the Curtain at GE," *New York Times,* March 6, 2009.

130 *"You couldn't help thinking":* Ibid.
"But no investor": Ibid.

131 *Two hours into the intense meeting:* Kate Haywood and Prabha Natarajan, "GE Execs Maintain Optimistic Tone at Investor Mtg," *Dow Jones Newswires,* March 19, 2009.
"sense-making project": Comstock, *Imagine It Forward,* pp. 242, 243.

21. KABLETOWN

132 *its declining viewership:* Sam Schechner, Jeffrey McCracken, and Max Colchester, "Comcast, GE Ready to Announce Deal," *Wall Street Journal,* December 3, 2009.

133 *Broadcast was fading:* Ibid.

134 *The remaining stake:* Ibid.
valuation of the proposed deal: Bob Wright, with Diane Mermigas, *The Wright Stuff: From NBC to Autism Speaks* (New York: Rosetta Books, 2016), pp. 234–235. *As Wright saw it:* Ibid.
"I've never had": Jeffrey Immelt, "A Conversation with Jeff Immelt, the CEO and Chairman of General Electric," *The Charlie Rose Show,* November 8, 2007. *The NBC Universal division:* Paul Glader and Jeffrey McCracken, "The Comcast-NBC Talks: NBC Has Been Something of an Odd Fit with Its Parent," *Wall Street Journal,* October 2, 2008.

135 *GE had bought RCA:* Paul Glader, "GE, Looking to Shed NBC Unit, Refocuses Its Vision," *Wall Street Journal,* November 20, 2009.

22. NO OTHER WAY

136 *Jeff Immelt traveled:* Jeff Immelt, "Renewing American Leadership," speech delivered at the US Military Academy, West Point, NY, December 9, 2009.

139 *The OTS wasn't expected:* Damian Paletta, "Regulator Clash Shows Stakes on
 Eve of New Banking Rules," *Wall Street Journal,* June 5, 2009.
 Immelt was openly talking: Jeff Immelt, remarks at the Electrical Products
 Group Conference, Sarasota, Florida, May 19, 2009.
140 *Capital begrudgingly accepted:* Michael Neal, remarks at Sanford C. Bernstein
 & Co. Strategic Decisions Conference, New York, May 31, 2012.
 "going to be with us": Ibid.

23. GREEN IS GREEN

142 *The cost of disposing:* Robert Sussman, EPA Senior Policy Counsel, email to
 Walter Mugdan et al. ("GE Feedback from Brackett Dennison"), December 8,
 2010, https://www.foiaonline.gov/foiaonline/api/request/downloadFile/34_ED
 _003002_00000110.pdf/7bb0ecc5-5463-4d59-846b-832c32e40f44.
143 *The lowlight came:* John Nolan, "GE Defends Its Record on PCB Contamina-
 tion," *Associated Press,* April 22, 1998.
146 *"Imagine what twenty years":* Bill Moyers, "The Fight to Save the Hudson
 River," in *America's First River: Bill Moyers on the Hudson,* April 24, 2002,
 https://billmoyers.com/content/americas-first-river-bill-moyers-on-the
 -hudson-the-fight-to-save-the-river/.
 "Why are they afraid": New Jersey General Assembly Solid and Hazardous
 Waste Committee, "Committee Meeting: Testimony from Concerned Public
 Policy Makers and Environmental Experts on the Appropriate Role GE
 Should Play in the Cleanup of the Hudson River," New Jersey Office of Leg-
 islative Services, Public Information Office, Hearing Unit, June 19, 2001, pp.
 51–52, https://www.njleg.state.nj.us/legislativepub/PUBHEAR/061901LB.pdf.
147 *"You're going to make us":* Comstock, *Imagine It Forward,* p. 135.
149 *"One thing is clear":* Joshua Ozersky, "Working in a Coal Mine: Lord, I Am So
 Tired, but Good-Looking," *New York Times,* July 3, 2005.
 weighed in that August: Daniel Fisher, "GE Turns Green," *Forbes,* August 15,
 2005.
150 *Now Jackson was calling:* Walter Mugdan, EPA Deputy Regional Administra-
 tor, email to David King et al. ("FYI — Re Contacts with GE"), December 16,
 2010, https://www.foiaonline.gov/foiaonline/api/request/downloadFile/32_ED
 _003002_00000108.pdf/61a08284-cf90-41b8-a3f8-bcee1ba61a8f.

24. PALATE CLEANSER

152 *convince multiple news outlets:* Devin Leonard and Rick Clough, "How GE
 Exorcised the Ghost of Jack Welch to Become a 124-Year-Old Startup," *Bloom-
 berg Businessweek,* March 17, 2016; Steve Lohr, "GE, the 124-Year-Old Software
 Start-Up," *New York Times,* August 27, 2016.
153 *"The engineers have to recognize":* David Kirkpatrick, "GE's Comstock: The
 Imperative Is Speed," *Techonomy,* November 21, 2014.

154 *a second-time entrepreneur:* Ben Kaufman, "Behind the Invention: The Mophie Juice Pack," *Medium*, December 31, 2013.

27. PLAYING STARTUP

171 *By March of 2016:* Leonard and Clough, "How GE Exorcised the Ghost of Jack Welch to Become a 124-Year-Old Startup." *Bloomberg Businessweek*, March 17, 2016.

29. PROJECT HUBBLE

179 *Such thinking had been inspired:* John C. Coates, John D. Dionne, and David S. Scharfstein, "GE Capital After the Crisis," Harvard Business School Case 9-217-071, May 3, 2017.

31. BIG BETS

188 *"software is eating":* Marc Andreessen, "Why Software Is Eating the World," *Wall Street Journal*, August 20, 2011.

189 *this revelation:* Steve Lohr, "GE, the 124-Year-Old Software Start-Up," *New York Times,* August 27, 2016.
 Immelt proclaimed: Steve Lohr, "Security Risks in the Internet of Industry," *New York Times,* October 15, 2015.
 Its claim that: Ted Mann, "GE Boosts Pay to Land Tech Talent," *Wall Street Journal,* September 21, 2016.

190 *complex had inspired the setting:* Donna Larcen, "Corporate Types Identify with 'Dilbert' Humor," *Hartford Courant,* March 29, 1995.

32. DINNER IN PARIS

193 *Kron had already wooed:* Michael Stothard, "How Paris Repelled General Electric from Alstom Takeover," *Financial Times,* June 16, 2014.
 Bouygues had owned: Inti Landauro, "Bouygues Writes Down Alstom Stake," *Wall Street Journal Europe,* February 18, 2014.

33. ONE DAY IN CHICAGO

197 *The protesters' signs:* Communications Workers of America, "Standing Up for Retiree Healthcare," *CWA e-Newsletter,* May 1, 2014, https://cwa-union.org/news/entry/cwa_e-newsletter_may_1_2014.

34. A MODEST PROPOSAL

202 *He'd considered Areva:* David Gauthier-Villars and Stacy Meichtry, "French CEO Defies Paris, Spurred by His Past," *Wall Street Journal,* May 1, 2014.

Kron had taken the meeting: Stothard, "How Paris Repelled General Electric from Alstom Takeover."

203 *pronouncing the GE deal:* Inti Landauro, "French Minister Wants GE to Change Offer for Alstom Unit," *Dow Jones Newswires*, May 20, 2014.

204 *"breach of national ethics":* Gauthier-Villars and Meichtry, "French CEO Defies Paris, Spurred by His Past."

35. "WE KNOW WHAT WE'RE DOING"

206 *arrived in June:* Inti Landauro and Stacy Meichtry, "Alstom Suitors Dig In for a Siege," *Wall Street Journal*, June 19, 2014.

207 *"All of the operators":* Stacy Meichtry and Inti Landauro, "France Urges EU to Relax Antitrust Rules," *Wall Street Journal*, June 27, 2014.

36. BUY HIGH

209 *the oil and gas equipment market:* Ted Mann, "GE Chief's Overhaul Slips on Oil — Plunging Crude Prices Crimp Immelt's Long-Awaited Remake of the Conglomerate He Inherited," *Wall Street Journal*, March 4, 2015.

211 *"base case" assumption:* Ibid.

213 *what happened to big plans:* Steve Knight, "GE Pulls Plug on Historic Lufkin Foundry," *Longview News-Journal*, August 24, 2015.

 wildcatters had it worse: Nicole Friedman, "Oil Sinks Below $40 Amid New Signs of Glut," *Wall Street Journal*, December 3, 2015.

37. THE PROBLEM WITH OWEN'S HAMMER

216 *"industrial strength platform":* General Electric, "GE Launches 14 New Industrial Internet Predictivity Technologies to Improve Outcomes for Aviation, Oil & Gas, Transportation, Healthcare, and Energy," press release, October 9, 2013.

217 *integrating a software operation:* Mann, "GE Boosts Pay to Land Tech Talent."

 "People inside the company": Joe Lazauskas, "This Content Campaign Increased Applications to Work at GE by 800 Percent," *Contently*, June 20, 2016.

218 *half the R&D budget:* Gregory Hayes, President and CEO of United Technologies Corporation, presentation at Barclays Industrial Select Conference, Miami Beach, Florida, February 18, 2015.

38. "I DON'T EVEN KNOW WHAT WE'RE SELLING"

223 *the conglomeration of Predix's code:* Alwyn Scott, "GE Shifts Strategy, Financial Targets for Digital Business After Missteps," *Reuters*, August 28, 2017.

39. JEFF'S DEAL

226 *Justice Department had zeroed in:* For the Justice Department suit filed in Sep-

tember 2015 against GE, Alstom, and PSM, see https://www.justice.gov/atr/case
-document/file/768341/download. The government's argument was that the
combination would "give GE the ability to raise prices or decrease the quality of
service provided to . . . customers. In addition, the proposed acquisition would
eliminate PSM as a vigorous product innovator for the GE installed base and
likely would reduce GE's incentive to innovate in response to PSM."

228 *the company had paid:* Brent Kendall and Joel Schectman, "Alstom to Pay
$772 Million to Settle Bribery Charges," *Dow Jones Newswires,* December 22,
2014.

 A GE spokesman: Ibid.

 "Put it this way": Ibid.

41. AN UNEXPECTED GUEST

236 *"It's not something":* David Benoit and Ted Mann, "Activist Investor Makes Big
Bet on GE," *Wall Street Journal,* October 5, 2015.

42. THE CANDY FACTORY

238 *Fairfield as "a morgue":* Thomas Gryta, "What's Behind GE's Move from the
Connecticut Suburbs to Boston," *Wall Street Journal,* May 15, 2017.

240 *"In a complicated world":* General Electric 2016 annual report.

 In the annual letter: Ibid.

241 *"We have delivered":* Ibid.

 the first natural gas turbine: Sonal Patel, "A Brief History of GE Gas Turbines,"
Power, July 8, 2019.

 He had been running: Kate Linebaugh and Joann S. Lublin, "For Retiring
GE Executive, $89,000/Month Not to Work," *Wall Street Journal,* August 2,
2012.

43. YOUR PEOPLE DON'T WANT IT BAD ENOUGH

243 *"What is Predix":* Steve Bolze, speaking at JPMorgan Aviation, Transporta-
tion, and Industrials Conference, New York, March 9, 2016.

244 *"The world needs":* Ibid.

 "So there will be growth": Josef Kaeser, speaking at Bank of America Merrill
Lynch Conference, London, March 18, 2016.

44. MAKING ADJUSTMENTS

249 *move would catch:* Thomas Gryta, "GE Shares Tumble amid Big Payout Cut,
Criminal Probe," *Wall Street Journal,* October 31, 2018.

46. MANAGING POWER

257 *company's tax assets included:* Ted Mann, "GE Nears Decision on Relocating

Its Headquarters," *Wall Street Journal*, September 10, 2015, https://www
.wsj.com/articles/ge-nears-decision-on-relocating-its-headquarters
-1441931084.
"We don't need trade deals": General Electric 2016 annual report.

47. UNMOVABLE TARGET

260 *"We didn't want to"*: Jeff Immelt, remarks at the Electrical Products Group
Conference, Sarasota, FL, May 24, 2017.
261 *Ed Garden had recently made*: David Benoit, "How Trian Thinks About GE's
Cost Cuts — Market Talk," *Dow Jones Newswires*, April 27, 2017.

48. THE NEW GUY

263 *chose the location*: Steve Lohr, "GE, Pressured by Its Investors, Changes
Leader," *New York Times*, June 13, 2017.
264 *Friday in early June*: Ibid.
266 *"I owe myself"*: John Flannery, General Electric CEO succession investor
meeting, June 12, 2017.

49. BEHIND THE CURTAIN

268 *Montana political blogger*: Montana Cowgirl, "Top Ten Notable Events at the
Butte Economic Forum," September 15, 2010.
269 *Immelt stepped down*: Thomas Gryta and Joann S. Lublin, "New Chief at GE
Starts Undoing the Costly Past," *Wall Street Journal*, October 19, 2017.
271 *grew up in Cleveland*: Maria Saporta, "GE Energy's Russell Stokes Powers Up
in Atlanta," *Atlanta Business Chronicle*, January 27, 2017.
272 *with just ten days left*: Jeff Immelt, General Electric second-quarter conference
call, July 21, 2017.
"There is both an opportunity": Ibid.
In what seemed like a parting shot: Ibid.

50. TRIAGE

275 *Before officially stepping down*: Greg Bensinger and Thomas Gryta, "Uber CEO
Search Takes New Twist," *Wall Street Journal*, August 17, 2017.

51. CLEANING HOUSE

280 *Comstock would later say*: Stephanie Clifford, "Beth Comstock on How to
Survive Losing Your Job," *Marie Claire*, January 15, 2019.
282 *"Nelson Peltz is a genius"*: Joshua Brown, Twitter tweet, posted July 17,
2017.
Cornell-educated engineer: Verizon, "Executive Bios: Lowell C. McAdam,"
https://www.verizon.com/about/our-company/lowell-c-mcadam.

52. CONSERVATIVELY CALCULATED

286 *Buffett and his longtime business partner:* Warren Buffett, Berkshire Hathaway shareholder letter, February 25, 2012.
It's the second requirement: Ibid.

54. A BILL COMES DUE

294 *Justice Department investigation:* Thomas Gryta and David Benoit, "In GE Probe, Ex-Staffers Say Insurance Risks Were Ignored," *Wall Street Journal,* November 30, 2018. As of this writing, the SEC and the Justice Department are continuing their investigation of GE's accounting in its Power division and insurance holdings. GE has said that it is cooperating with the government investigators. GE has also denied allegations of fraud made in shareholder lawsuits, some of which have been dismissed.
senior executives in the division: Gryta and Benoit, "In GE Probe, Ex-Staffers Say Insurance Risks Were Ignored."

295 *Bornstein started talking:* Jeffrey Bornstein, remarks at Barclays Industrial Select Conference, Miami Beach, Florida, February 22, 2017.
more serious that July: Jeffrey Bornstein, General Electric second-quarter conference call, July 21, 2017.
For Jamie Miller: General Electric investor update, November 13, 2017.

55. ASLEEP AT THE SWITCH

297 *described the insurance situation:* General Electric insurance update call, January 16, 2018.

299 *seventeen independent directors:* General Electric proxy statement, March 12, 2018.

56. "MANAGING IN A BROADER SENSE"

301 *Flannery had stunned the room:* Joann S. Lublin and Kate Linebaugh, "Meet the Next CEO of General Electric: John Flannery," *Wall Street Journal,* June 12, 2017.

302 *"So being deliberate":* John Flannery, remarks at Electrical Products Group Conference, Boca Raton, FL, May 23, 2018.

57. A MIDSTREAM CHANGE

307 *Federal criminal and civil investigators:* Gryta and Benoit, "In GE Probe, Ex-Staffers Say Insurance Risks Were Ignored."

309 *"It will take years":* Jeff Immelt, "How I Remade GE," *Harvard Business Review,* September/October 2017.

BIBLIOGRAPHY

Bair, Sheila. *Bull by the Horns: Fighting to Save Main Street from Wall Street and Wall Street from Itself.* New York: Free Press, 2012.

Blackwelder, Julia Kirk. *Electric City: General Electric in Schenectady.* College Station: Texas A&M University Press, 2014.

Comstock, Beth. *Imagine It Forward: Courage, Creativity, and the Power of Change.* New York: Crown/Currency, 2018.

Cowie, Jefferson. *Capital Moves: RCA's Seventy-Year Quest for Cheap Labor.* New York: New Press, 1999.

Evans, Thomas W. *The Education of Ronald Reagan: The General Electric Years and the Untold Story of His Conversion to Conservatism.* New York: Columbia University Press, 2006.

Hammond, John Winthrop. *Men and Volts: The Story of General Electric.* Philadelphia: J. B. Lippincott Co., 1941.

Israel, Paul. *Edison: A Life of Invention.* New York: John Wiley & Sons, 2000.

Kennedy, Rick. *GE Aviation: 100 Years of Reimagining Flight.* Wilmington, OH: Orange Frazer Press, 2019.

Lane, Bill. *Jacked Up: The Inside Story of How Jack Welch Talked GE into Becoming the World's Greatest Company.* New York: McGraw-Hill, 2008.

Langone, Ken. *I Love Capitalism! An American Story.* New York: Portfolio/Penguin, 2018.

Magee, David. *Jeff Immelt and the New GE Way: Innovation, Transformation, and Winning in the 21st Century.* New York: McGraw-Hill, 2009.

Martin, James. *In Good Company: The Fast Track from the Corporate World to Poverty, Chastity, and Obedience.* Lanham, MD: Sheed & Ward, 2000.

O'Boyle, Thomas F. *At Any Cost: Jack Welch, General Electric, and the Pursuit of Profit.* New York: Alfred A. Knopf, 1999.

Paulson, Henry M., Jr. *On the Brink: Inside the Race to Stop the Collapse of the Global Financial System.* New York: Business Plus, 2010.

Ries, Eric. *The Lean Startup: How Today's Entrepreneurs Use Continuous Innovation to Create Radically Successful Businesses.* New York: Crown Business, 2011.

Sorkin, Andrew Ross. *Too Big to Fail: The Inside Story of How Wall Street and Washington Fought to Save the Financial System — and Themselves.* New York: Viking Penguin, 2009.

Tichy, Noel M., and Stratford Sherman. *Control Your Destiny or Someone Else Will.* New York: HarperCollins, 1993.

Vonnegut, Kurt. *Player Piano.* New York: Delacorte Press, 1952.

Welch, Jack, with John A. Byrne. *Jack: Straight from the Gut.* New York: Warner Business Books, 2001.

Welch, Jack, with Suzy Welch. *Winning.* New York: HarperCollins, 2005.

Wright, Bob, with Diane Mermigas. *The Wright Stuff: From NBC to Autism Speaks.* New York: Rosetta Books, 2016.

Documents

New Jersey Legislature, General Assembly, Solid and Hazardous Waste Committee, *Testimony from Concerned Public Policy Makers and Environmental Experts on the Appropriate Role GE Should Play in the Cleanup of the Hudson River.* Englewood, NJ: June 19, 2001.

INDEX

ABOUT THE AUTHORS

Ted Mann is a reporter in the *Wall Street Journal*'s Washington bureau, where he is part of a team covering business and government. He was the beat reporter covering General Electric and other industrial conglomerates for the corporate bureau in New York from 2014 to 2017, and he previously covered transportation for the Greater New York section. Before joining the *Journal*, he covered politics and the Connecticut state government for *The Day* of New London. A graduate of New York University, he lives in Washington with his wife and son.

Thomas Gryta, a reporter in New York for the *Wall Street Journal*, covers General Electric and previously covered the American telecommunications industry. Before spending a year at Columbia University as a Knight-Bagehot Fellow, he covered the biotechnology and pharmaceutical industries. He has worked as a copy editor and breaking news reporter, after breaking into journalism as a news assistant in London. He studied history at the University of Massachusetts–Amherst and spent a year studying in Germany. He lives in New Jersey with his wife and children.